Expe[illegible]ng with Nu[illegible]

A GUIDE FOR PRESCHOOL, KINDERGARTEN, AND FIRST GRADE TEACHERS

Margaret Stern

Illustrated by John Roberts

Acknowledgements

...ne Stern ... g those years I It has been ... ing and have published *Exper*... reading. Hence, this have gathe... the earlier work, but as a ...tor, I have kept the original title. ...k the Carnagie Corporation of New York for giving the Structural Arithmetic Program two grants in 1958 and 1959 and report that their generosity has continued to promote the original intention of those grants for five decades.

I especially want to thank Elizabeth Freidus, founder of the Gateway School of New York. She encouraged me to carry out experiments with children with learning disabilities and to discover how teachers learn. I thank the staff for their energy and time, and their head teacher, Tonya Pulanco, for her contribution of many ingenious new experiments.

For understanding my work and giving me help when I needed it, I want to thank Peggy Mabie, director of her own educational center in Port Ewen, New York.

Though I have written this book alone, I have continued to enjoy the collaboration of my coauthor, Toni Gould, as we work on the revision of the Structural Arithmetic Series.

I gratefully acknowledge the encouragement and initiative of my publisher and give thanks to Diane Winkleby for her editorial assistance and creative suggestions, and to Helaine Randerson for her accurate and efficient typing job.

I am most grateful to Fritz Stern who believed in my work from the beginning, and to my children, Fred and Katherine, who always enjoy teaching with their grandmother's blocks.

The Author

Margaret B. Stern is a graduate of Wellesley College and has a master's degree from the Bank Street School of Education. She has spent much of her time training math teachers and teachers of children with learning problems, has lectured extensively, and conducted workshops throughout the United States and overseas. She has been a special lecturer at many universities, including Teacher's College, Columbia University, New York University, University of Vermont, and New Jersey State Teacher's College. Margaret Stern is coauthor of the *Structural Reading Program*. Her other publications include *Gould-Stern Early Reading Activities*, *Children Discover Arithmetic*, and the *Structural Arithmetic Program*.

The Artist

John Roberts is a graduate of the Rhode Island School of Design; he earned his master's degree from the Rochester Institute of Technology. He is a deaf artist who also taught in the public schools of Vermont. His exceptionally clever illustrations should please teachers everywhere and teach them how to enjoy children's enthusiasm for learning mathematics.

Printed in U.S.A.
ISBN 0-9779132-4-4

Contents

Introduction

When we teach mathematical truths and facts by rote, we take away from children not only the joy of using their minds but also their sense of independence. They feel manipulated, and the result is unimaginative mechanical work. In some cases this leads to boredom; in others, to a turning off of the mind, to withdrawal, and to failure. What children crave is freedom to carry out their own experiments and to draw their own conclusions. To the degree that they can do this, their work will be fulfilling; they will develop a sense of pride in themselves and in their ability to think for themselves.

The Structural Arithmetic materials are designed so as to enable children to learn on their own. The experiments and games gathered together in this new edition of *Experimenting with Numbers* are the result of the further development of the original work published by Catherine Stern (Houghton Mifflin, 1950). They have been tried out successfully in classrooms for over thirty-five years. Teachers have expressed their appreciation for the many illustrations that make playing the games so much easier for them.

The first experiments take place in the counting board, where children fit number blocks into matching grooves. They discover that each number has a place of its own in the series from 1 to 10, and a different relationship to each of the other numbers. In the 10-box, they fit together pairs of number blocks that total 10. Once they learn the name of each block they can discover for themselves the combinations that make 10. By filling smaller and smaller boxes with blocks, they figure out the combinations with sums less than 10. The pattern boards, on the other hand, reveal quite different concepts. By filling these boards with colored cubes, children see familiar facts in a new light based on the characteristics of odd and even numbers. Children discover that the ultimate authority in learning mathematics lies with these experiments that they can perform on their own.

This approach is exciting for children of all abilities. It allows them to make great leaps forward in mathematical thinking. Furthermore, the number concepts they develop are sound from the very beginning. They need not discard these early formulations as deceptively simple but can carry them forward as the building blocks of algebra.

Abilities Needed in Forming Concepts

For children with learning disabilities, these materials are of special importance. These children have difficulty in discrimination, memory, intersensory organization, perceptual processing (both visual and auditory), and the ability to sustain their attention. Children may be struggling with a combination of these deficits. They must learn how to receive and to integrate information from as many different senses as possible in order to form concepts. They have difficulty in "getting things from books," but trust their ability to learn when they have real experiences. Structural Arithmetic allows children's thinking to grow out of their own experimenting with materials.

Teachers are eager to know why certain children fail, where the trouble comes from, and how they can help them. They realize they must observe the children's reactions closely and must be able to move quickly to something less threatening if a new step should prove too difficult. They need to know how to create situations in which these pupils can be sure of success. Suggestions of this kind are given at the point teachers most need them, in the directions accompanying each experiment.

In *Experimenting with Numbers* children progress from games with concrete materials in which they form basic concepts to the recording of addition and subtraction facts with symbols. Aiding in the formation of concepts are some important factors, which are highlighted throughout the book: visualization, comprehension of language, auditory memory, the use of action in learning, and the ability to reverse an idea. *Experimenting with Numbers* shows how each of these contributes to the unfolding of the children's ability to conceptualize and how the materials aid the children in developing the ability to think for themselves.

What Is Involved in the Formation of Concepts?

Visualization and the Formation of Concepts

Children use mental pictures with which to reason. Thus, in teaching children to think, we must develop their ability to form images. The materials will have fulfilled their purpose only when the children can visualize them without their being present. However, it is not sufficient for children to picture a quantity in isolation. They must be able to visualize it in relation to other numbers and to visualize the actions that can be performed on it. For instance, they must know what happens when a number is increased and, inversely, when the new total is diminished.

How can this ability to visualize be developed? The games and activities in *Experimenting with Numbers* present a variety of approaches that have proved themselves in many different classrooms.

Any hiding game delights children and presents a vast number of opportunities to improve mental imaging. In a hiding game, part of the model is hidden while the surrounding context is left visible. For example, have the children build a stair with the blocks from 1 to 10. Then ask them to close their eyes while you remove one step, perhaps the 5-block. When they open their eyes, they will see a gap in the stair between the fourth and sixth steps that helps them visualize the size of the missing block. They can verify

Structural Arithmetic Materials Needed for *Experimenting with Numbers*

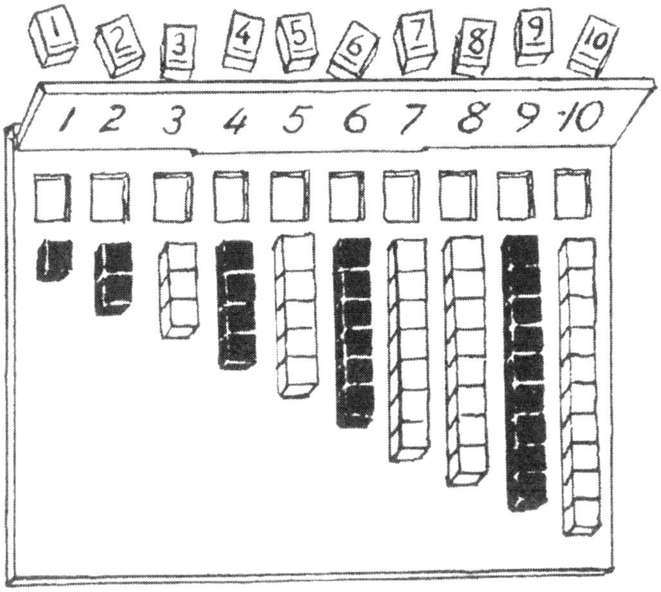

Counting Board with Number Markers 1 to 10, Number Guide, and Number Blocks

Extra Number Markers and Signs

Number Boxes 1 to 10

10-Box with Number Blocks

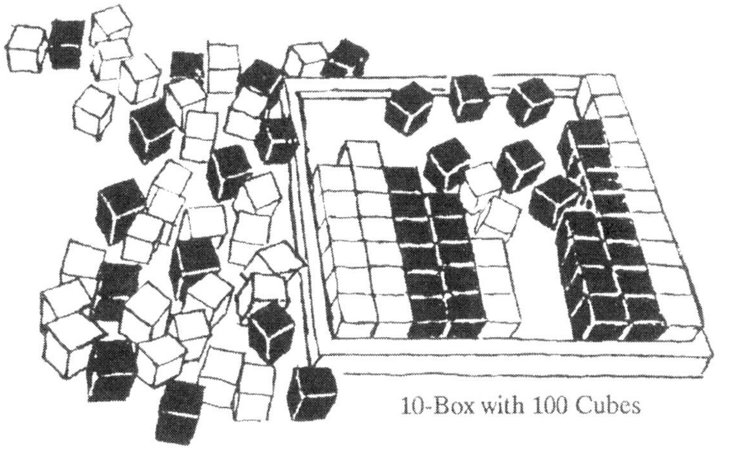

10-Box with 100 Cubes

Pattern Boards 1 to 10

their mental image immediately. From among the blocks scattered on the table, they pick out the 5-block and reinsert it into the stair. Their concept of 5 has been enlarged. They know where 5 comes in the stair and the way in which 5 is related to 4 and to 6 (see page 19, #5).

Language and the Formation of Concepts

The best way to teach the meaning of spoken language is to give children the opportunity to see and touch what the words describe. Presenting children with models to handle enables them to work out for themselves what the spoken words mean. For example, when children hear an order such as "Find the number that is one bigger than 5," they know they can figure it out with the blocks. They verify the words immediately by placing the 6-block after the 5-block and feeling that it is one unit bigger. Language becomes alive with meaning from the very first word. Such insights are not easily gained when children use the piecemeal approach of counting or depend on printed number symbols that do not show the meaning of "one bigger than."

Language is a vehicle for thought. Pupils should have many opportunities to put new concepts into their own words. Parroting the words of adults is not true learning; words not understood are easily forgotten. Rather, an exciting moment occurs when children use their own vocabulary to describe a concept they understand. Furthermore, children love to take the role of teacher. For example, while the other children's eyes are closed, a leader selects two blocks that make 10 (the 8-block and the 2-block) and hides them. The children open their eyes and watch as the leader reveals one of the number blocks (8) and places it in the 10-box. The leader calls on a friend and asks a question such as "Which block fits in here with 8 to make 10?" After the friend answers, the leader verifies this answer by putting the 2-block above the 8-block to make a total of 10 (see page 39, #9). By listening to the language children use, the teacher can judge their stages of conceptual development. The child in the example was able to communicate a difficult concept satisfactorily. It is generally known that trying to teach clarifies one's thinking.

Receptive Language, Auditory Memory, and the Formation of Concepts

In carrying out spoken directions, children develop their skills in comprehending receptive language. It is fun for them to follow directions that are part of a game. For example, scatter blocks 1 to 10 on a table across the room. Ask a child to go to the table and get 5 and the block one smaller than 5. This gives the teacher a way to test children's auditory memory and language comprehension. They should return and place the 5-block and the 4-block into the counting board (see page 25, #6). The other children are reassured when a classmate can follow directions perfectly. Developing receptive language and a good auditory memory is especially important for children whose language is immature.

Action and the Formation of Concepts

Piaget explains that mathematical concepts develop as a result of the *actions* children perform *with objects,* not from the objects themselves. This should make it clear why pictures of objects in workbooks don't teach children. The Structural Arithmetic blocks cry out to be handled, thus enticing children to perform actions at every step--actions that help them develop mathematical concepts. What follows is an experiment in which the teacher encourages the children to visualize an action. The teacher displays a pattern of 4 cubes and asks the children to name the amount. The teacher then hides the pattern of 4 cubes behind a screen and "does something" to it. In a few seconds, the teacher reveals a changed pattern (now 2 cubes) and asks, "What did I do to it? How many are left?" The children study the new pattern and answer, "You took 2 cubes away; 2 are left." To say this, they must have been able to reconstruct in their minds what took place behind the screen--the subtraction of 2 cubes. Experiments like this give children practice in visualizing an action or an operation. By eliciting information, the teacher helps the children conceptualize the operation that took place (see page 54, #11).

Reversibility and the Formation of Concepts

The Structural Arithmetic program presents concepts in such a way that children can always figure out how they can "undo" what has been "done." For example, children see that adding 1 cube to a pattern of 4 cubes yields the next higher number pattern. Right away they learn to reverse this. They subtract 1 cube from the 5-pattern and find that this takes them right back to the original 4-pattern (see page 30, #3 and #4).

In a different experiment, children learn that the 6-block and the 4-block together measure 10. They find it entirely logical then that if they remove the 6-block from this combination, the 4-block must remain (see page 58, #5). From the beginning they understand that subtraction "undoes" what addition has "done." They realize that addition and subtraction have a special relationship to each other. Children must be able to turn an idea around in their minds before they can make adequate use of it.

Recording Mathematics

Mathematics is a mental activity--thinking with numbers. What part do symbolic processes play in this creative process? Writing numbers on paper is a way of clarifying ideas and setting them down. It is a more permanent form of thinking aloud. Once children can put their discoveries into words, they are ready to preserve them by recording them with the signs and symbols of mathematics.

In the Structural Arithmetic program each numeral is introduced as standing for a block of a certain length, the groove of corresponding size or the number of cubes it holds, or the pattern of a certain configuration. The pupils can see for themselves that an amount such as 7 has properties that are characteristic only of 7. For example, 7 is larger than the numbers preceding it and smaller than the numbers following it; it is an odd number; it can be constructed, or composed, of the subgroups 1 and 6, 2 and 5, and 3 and 4. Since the blocks and patterns are true representations of numbers, children can discover all these characteristics on a concrete level by themselves. Finally, the number symbols acquire all the qualities of the blocks and patterns for which they stand.

The Structural Arithmetic materials provide a plastic tablet, or number marker, for each number and symbol. Children can thus record equations before they have the coordination needed for writing them. It is less offensive for pupils to exchange one tablet for another than to erase what they have written. For example, the teacher places 7 cubes in the pattern board for 7. The children record this with the tablet for 7. The teacher then "does something" to the pattern of 7 cubes behind a screen and reveals the remainder--6 cubes. The children say, "You took 1 cube away." They select the tablets showing minus 1 and place them after the figure 7. The recorded operation looks like this: $7 - 1 = 6$ (see page 54, #12).

The opposite task is reading an equation. This is most fun when the equation on a card tells the children where to look for a reward such as a sticker. If, for example, the equation on the card is $7 + 1 = \_$, the child who knows the answer looks beneath pattern board 8 to claim the prize. Thus, the recording of concepts leads to the understanding of written equations (see page 50, #3).

Diagnostic Tools

The Structural Arithmetic materials make excellent diagnostic tools for assessing the strength of many different abilities, different aspects of which show up in these experiments. By observing the way children perform the tasks described in this book, teachers can assess the following abilities: to visualize, to comprehend spoken language, to put ideas into words, to remember names, to describe a category, to learn from sensorimotor experiences, to turn concepts around in the mind, to remember symbols, to reason and make inferences, and to develop the coordination required for recording numbers and equations.

These structured materials encourage children to handle them, to make choices, and to discover mathematical ideas. Children learn to formulate the discoveries they have made in the language of mathematics. By working out these relationships, they can see the part they play in the ordering of their experiences. They become confident as they experience success at the most basic levels of learning.

Suggestions for the Classroom

Teaching Children in Small Groups

The teacher who is able to develop the trust and friendship of a small group of children will gain powerful allies to work with. The games and activities in *Experimenting with Numbers* have been prepared with this kind of teaching in mind.

Learning from Discussion

Discussion among children can be an important aspect of their learning. It gives them an opportunity to clarify their ideas by putting them into words. Listening to others helps them see things from another point of view, while others' thoughts may stimulate them to form new ideas. Benefits from these exchanges between children are so great that teachers will find it worthwhile to develop the skills needed to create the warm atmosphere conducive to learning in a group.

Fostering Concentration

In any group there are some easily distracted children who need to be freed from distraction before they can concentrate on the task at hand. A basic rule of behavior offers special protection and help for some children--"Put your hands on the edge of the table." Once their hands are at rest, the children can listen to the teacher, focus on the game, and be ready for their turns. This behavior is asked of the whole group, and their response as a group enables the teacher to avoid humiliating one or two errant children. Since the others have not heard the teacher reprimanding certain individuals, they are less likely to single them out later to pick on. Children sense that the teacher's attitude is one of understanding and kindness. They become less anxious about how they will be treated when they slip from the standards of good behavior. As soon as the atmosphere has become more relaxed, they begin to pay better attention.

Keeping Children Involved in Learning

Children who have difficulty sustaining their concentration need to be reminded to pay attention often. It is easy to spot the children who think, "Oh, it's not my turn," because they will have stopped paying attention. Usually they do not realize that they can learn from watching others learn. The teacher should keep them involved in learning during another child's turn by asking questions. "Do you know what I just said to Chris?" or "Can you help Mary by giving her time to think?" As soon as children realize that an attentive attitude is a help to the whole group, they stay involved; they begin to respond to the needs of the others.

Children's errors are very often part of genuine experimentation and should be treated that way. It is not necessary to humiliate children. When teachers say "No!" or "Wrong!" they cause children to withdraw and wait for the authority to tell them what to do next. A better approach is to encourage them to try to deal with the situation. They learn best when they figure out something for themselves. Teachers will find suggestions on how to do this in the instructions for the games that follow.

Reorganizing a situation so that many children can participate is another way to keep them involved. Instead of one child's performing a task, such as building a number pattern with cubes, each child can build his or her own pattern (see page 29, #2). An attractive new job is created by this change; a leader must be appointed to check the others' patterns. Work organized in this way keeps all the children involved.

Fostering Cooperation

Children understand that games need rules. They sense that the rules are necessary for protection--perhaps from dominating individuals. All the children want to be part of the group. Here are some suggestions to foster their cooperation. "I'm watching until everyone is ready." Later, attention can be coaxed with, "Where are your controls?" or, to quell restlessness during another's turn, "Look at your friend. She is trying to think; are you helping?" Often children are not aware of making a noise or even that it is bothering anyone. The teacher can politely say, "Ssh! He needs to hear what I am saying," or at another time, sadly, "You're not ready! I'll have to skip you." But always reassure the child, "I'll come back to you when you are ready!" Teachers build self-respect when they give children good reasons for restraining themselves. The result should be a willingness to take turns, to listen, and to cooperate.

Building Psychological Support

One of the most important advantages of working in a group is the constant opportunity to observe others learning. Discouraged children have a chance to observe other children struggling with similar problems. As passive observers, not under pressure to perform, they can think more clearly about another's problems. When watching others succeed, they take heart, hoping that some day this will certainly happen to them, too. The group can give its members tremendous psychological support. It is the respect and concern given a struggling child by his or her surrounding friends that helps that child stay with the task, to persevere and work it out. The teacher can then say, "See, you did it! Good!" The children learn to respect their ability to work something out on their own.

Goals and Objectives: Level I

Number Games without Number Names

The design and order found in the Structural Arithmetic materials provide children with excellent conditions for learning about mathematical reasoning. From the very beginning, children are filled with curiosity about how to use the materials. They compare blocks to fit grooves, put colored cubes into pattern boards, build sequences from 1 to 10, and fit together blocks that enable them to discover the combinations that make 10. On Level I they learn to do the following tasks:

The Counting Board (no number names yet)

- To put each block in the groove where it fits
- To find the block for the groove you indicate
- To realize when a block is too small or too big for the groove and to find the one that fits
- To know where each block "lives" in the sequence

The Pattern Boards and Cubes (no number names yet)

- To recognize the matching pattern board for any cube pattern you build
- To be able to build the pattern with cubes for any board you show them
- To build the sequence from the smallest to the largest amount
- To know where each pattern "lives" in the sequence

The 10-Box (no number names yet)

- To fit together the blocks so they form the combinations that fill the 10-box
- To build a stair with the blocks from the smallest to the largest
- To find the missing block when one step has been removed

The Counting Board: Level I

Introduction

The Counting Board and Number Blocks

The children carry out their most basic experiments as soon as they pick up the number blocks, compare them, and fit them into their matching grooves in the counting board. They discover that each block is a different size, and they can locate its own special place in the series of numbers from 1 to 10.

On Level I, the earliest level, sensorimotor learning takes place. Children measure and fit blocks into grooves in the counting board, thus experimenting with materials as though they were puzzles. They develop basic number concepts through seeing and feeling number blocks of different lengths. Then on Level II they learn the number names *one, two,* and *three,* and later, on Level III, the symbol that stands for each amount or number block, such as 1, 2, and 3.

Experimenting Comes Before Language

The experiments and games should be played with enjoyment. The teacher should do very little talking in the beginning. This helps children who are struggling with language because it allows them to focus on what they are doing. The children should be concentrating on learning concepts through handling the materials and experimenting with them. Once they are competent on Level I, the "doing" level, children will want to move ahead and learn to put their discoveries into words. On Level II, new games and experiments are described so the teacher can teach the children number names and the language of mathematics.

Developing Skills

Many different skills are practiced in these games. Most of the skills that are used in the development of number concepts are also basic to the ability to read.

a. *Eye-hand coordination* improves because children practice over and over again, aiming each block so as to fit it into its proper groove.

b. *The ability to scan* is used when children search for one block among many scattered blocks.

c. *The ability to judge sizes* is developed when children constantly compare blocks with empty grooves to find the matching combination.

Cognitive Growth

Teachers should allow children to work at a leisurely pace--the pace they need to absorb what they are doing. The teacher can sit back and enjoy watching them learning. The children can do a task such as building a sequence as soon as they realize that each block is one unit longer than the preceding block. In other experiments they discover that the big blocks "live at the end" of the counting board (one day to be referred to as "the right side"), while the little blocks come at the beginning of the counting board (later to be thought of as "the left side," where one must always start).

Teachers should realize that amounts larger than 5 units are difficult to identify. This was known by early cultures. The Romans subdivided the numerals for amounts over five (VI, VII, VIII). Human beings can see a group of 5 or fewer items at a glance, but they need help in recognizing amounts greater than 5. The 10- and 9-blocks are easily recognized as the biggest and the next to biggest, and in the series of blocks in the counting board, as last and next to last. However, the 6-, 7-, and 8-blocks are more difficult. They show no subgroupings and their positions are in the center of the sequence of blocks. Teachers should take special care when dealing with them.

The children will find they can rely on themselves to discover where each block goes because the counting board presents only one groove in which each block fits exactly. No teacher needs to intrude. The materials encourage children to think for themselves. When a block is too big or too small for a groove, they can see and feel in what way it does not fit and try others until they are successful. These are basic developmental steps in cognitive growth. If some children have skipped these steps at an earlier age, experimenting with the materials is an excellent way to gain those experiences now.

1. Discovering Size Relationships

Note: The awareness of relationships plays a central role in productive thinking.

- Display the counting board filled with blocks 1 to 10.
- Give each child a block from the board. (If there are fewer than ten children, some will get two blocks.)
- Point to an easy groove, i.e., a small or large groove (leave grooves 6, 7, and 8 until last).
- * Ask, "Who has the block that fits here?" (If the block doesn't fit, ask, "Does someone else have it?")
- Continue in random order until the board is filled.

Note: Remind the children that everyone must be able to see all the blocks, so they must not hold them in their hands but lay them on the table and put their hands on the edge of the table. It will help the children's concentration if you keep this rule. To prevent distortion, position the counting board so it faces the child who is working with it

2. Discovering Size Relationships

- Display the counting board filled with blocks 1 to 10.
- Then scatter the blocks on the table.
- * Point to an empty groove, such as the 3-groove, and say, "Find the block that fits here."
- Call on a child.
- If his or her block is wrong, ask, "Does it fit? Try another block."
- Continue in random order until the board is filled with blocks.

Note: In this game, the children must scan many blocks while looking for a block that fits. Encourage the children to learn by experimenting. When a block doesn't fit, move it back and forth in the groove to show that it does not fit. The children learn by trying different blocks. Don't ask, "Is it too little or too big?" It is difficult for them to know whether you are talking about the block or the groove (the block is too little, the groove is too big). Simply ask, "Does it fit? Try another block." Later on you can ask about size.

*An asterisk indicates that this part of the experiment is illustrated.

3. Exploring Size Relationships

- Display the counting board filled with blocks 1 to 10.
- Distribute the blocks among the pupils or let them choose their blocks. (If there are fewer than ten children, give each several blocks.)
- Give first turns to children with small or large blocks.
- * Ask, "Where does your block go?"

Note: While hunting for the grooves where their blocks will fit exactly, the children will discover that the small blocks are on the left side, while the larger blocks are on the right. For children who need it, simplify the task by leaving blocks 6, 7, and 8 in place. This prevents failures and gives the children a chance to master the task.

4. Building the Sequence from 1 to 10

Learning about Directionality

Vocabulary: *next* and *after*

- Display the counting board filled with blocks 1 to 10.
- Remove the blocks and scatter them on the table.
- Ask for the blocks in sequence. This means going from left to right.
- Begin by putting the 1-block in the counting board.
- * Ask, "Which block comes next?" (or "Which block comes after this block?").
- Call on someone; continue until all the blocks are in place.

Note: The children will be moving from small to large, left to right. When you use the word *next* or *after,* teach the meaning by pointing first to the block in place and then to the groove that comes next. When you say *next* or *after this block,* the children see that your finger is moving to the right. This teaches directionality. Many children cannot cope with the meaning of the words *left* or *right* yet. These words are not important at this point.

5. Finding a Block of Equal Size

Vocabulary: *twin*

- Fill the counting board with blocks.
- Place a duplicate set of blocks on the table.
- Explain that each block on the table is the twin of a block in the board.
- Point to a block in the counting board.
- Ask, "Can you find the twin of this block? Put it on top." (This shows that the blocks are equal.)
- Continue until each block has its twin lying on top of it.
- Put the twin blocks back on the table, ready for the next game.

The Hiding Game

Explain how to play the Hiding Game. (When the children's eyes are shut, you will take a block from the counting board and hide it under a box.)

- Say, "Ready. Close your eyes!"
- Take a block from the counting board and hide it under the box.

* Say, "Open your eyes! Who can find the twin of the block I have hidden? Are you a good detective? What clues do you see?" (Point to the empty groove.)

- Call on a child to pick out the twin block and put it into the empty groove.
- Now the child lifts up the box and reveals the hidden block!

6. Strengthening Visual Memory

- Remove the blocks from the counting board and place them on a table across the room.
- Point to an empty groove, such as the 4-groove.
- Put a toy plastic animal above it to indicate which groove.

* Say, "Run to the table. Get the block that fits here."

- Continue until all the blocks are in the board.

Note: Everyone watches to see whether the child called on will return with the right block. If the block doesn't fit, have the child run back for the right block. If the task is too difficult, ask another child to go along. Children who have difficulty alert you to their need for special help. Before they cross the room, strengthen their visual perception with touch. Have them feel the groove with their forefingers. This helps them concentrate on the size they are to remember.

7. Teaching Children to Be Teacher

Being teacher is exciting. It is an important experience in learning. However, when you ask children to be teachers, you are asking them to take over a role they do not fully understand. Here are two ways to teach this new role.

Structuring the Role Step by Step

- Call on the child who will be teacher.
- Show how to display the counting board and scatter the blocks on the table.
- Say, "Call on a friend."
- Say, "Point to a groove you want your friend to fill."
- * Elicit a teacher-like command from the child by asking, "What do you want your friend to do?"
- The child should answer, "Find the block that lives here."
- Say, "Then ask your friend that question."

Note: The children's pleasure is unbelievable.

Giving the Children a Model Demonstration (not illustrated)

- Tell the children to watch you carefully so they can learn how to be teacher.
- Play one of the first counting board games, perhaps #2, Discovering Size Relationships.
- Ask each question clearly, such as, "Who has the block that fits here?"
- Pass the board to the child who will be teacher.
- Tell the child to do exactly what you did.

Note: There is less talking by the teacher in the second approach. The presentation is given as a whole; it works well. However, many teachers prefer the step-by-step approach. There are usually some children whose success is dependent on the teacher's sequencing the steps for them.

8. Checking Up: Self-Directed Individual Work

Before moving on, check the children's competence on the prelanguage level. They will enjoy showing you that they know where every block goes in the counting board.

Remove the blocks from the counting board. Invite different children to fill the board entirely by themselves. Children vary in how they handle this task: some proceed from small to large blocks, others from large to small ones. This seems to help them organize the task. They seem to know that this approach minimizes their chance of making an error.

When they have reached a more advanced stage, the children will be able to pick up any block and place it in the board directly where it "lives."

How Much Experimenting?

Experimenting is a good way of learning. Children sometimes make up experiments that puzzle the teacher. For instance, they might try all the blocks that don't fit in the 5-groove--first, every block that is too large, then every block that is too small. Finally, with great glee, they put in the fitting block. It seems like a joke. You have to decide when to appreciate experiments like this. Sometimes the children are wasting time. During group games you might tell them they are using up other people's turns even though the experiment is interesting.

Pattern Boards: Level I

Introduction

Pattern Boards 1 to 10 and Cubes

Each of the 10 pattern boards provides empty blanks in a set pattern from 1 to 10. Each blank holds one cube. Children learn to recognize the pattern of any number by building it with cubes. They check their pattern against the original by placing each cube into a corresponding blank square in the pattern board, leading them to recognize the one-to-one correspondence between each element in their cube pattern and the model they are copying. It helps them understand the concepts of *too many, too few,* and *as many as.*

These patterns give the children another perception of each number. For many children the patterns are easier to remember than the number blocks. At the beginning of Level I, children copy the patterns without counting. They not only see, but also feel, that each cube must have a blank to fit into. These actions and sensory impressions develop basic number concepts.

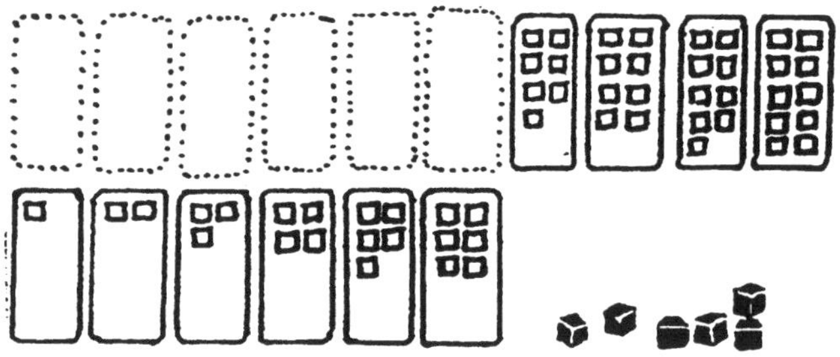

Studying Patterns 1 to 6

Since patterns 1 to 6 are easy to recognize, we study them first. The children learn that patterns 1, 3, and 5 can be built with the odd cube a different color.

This helps them understand how an odd number differs from an even one, and highlights the characteristics between numbers.

Practice in Left-to-Right Directionality

Children need much practice in going from left to right before learning to write letters and numbers. They get this essential practice if they build a pattern by putting in the left cube, then the right cube, and in the next line, "left, right." This shows them why the odd cube in patterns 1, 3, 5, 7, and 9 comes on the left side; it is the first cube in that row.

Constructing Cube Patterns: A Cross-Modal Task

It is difficult for certain children to look at a pattern board and then construct a like pattern out of cubes. Let us consider what is involved. This is a cross-modal task--a cross between visual and tactile modalities. The children look at the empty pattern board and study the pattern (visual). They try to build the same pattern on the table with cubes (tactile). They compare their pattern with the original one in the board (visual). They put the cubes into the pattern board (tactile). This task calls for the integration of information received from two different senses, sight and touch. When all the cubes fit, children know they have succeeded. Being able to complete such a task correctly is of great importance to all children.

The Display Card Helps Children Copy a Cube Pattern

Children with poor spatial organization find it almost impossible to copy a cube pattern on an open table top, suggesting how difficult it will be for them to copy letters and numbers on blank paper. They need guidepoints. An ingenious way to give them this guidance is to make display cards out of cardboard.

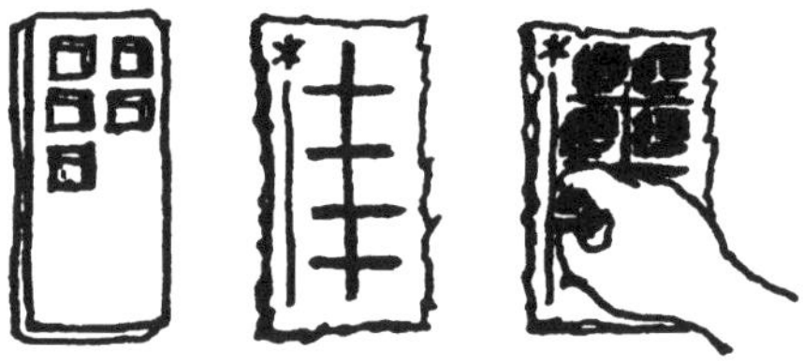

When the children look at the pattern board to be copied, they see a pattern framed by the edges of the board. We can echo this framework by making display cards for the children, giving them boundaries within which to orient themselves. This enables them to construct their own cube patterns. The display card has these other special features: a star in the upper left corner to show where to start, a colored line in the margin to indicate the left side, and a grid of lines to indicate the correct spacing of cubes. In time, the children will internalize these guidepoints and cease to need them.

1. Recognizing Number Patterns

Concepts: *too many, too few,* and *just right*

- Arrange pattern boards 1 to 6 in random order.
- Keep the display card and cubes for your use.
- Build a cube pattern such as the 4-pattern on your display card.

* Ask, "Where will this cube pattern fit?" (If necessary, say, "exactly fit?")

- Invite a child to check by putting the cubes into the board.
- Let each child keep a board for the next game.

Note: Visual cues are backed up by tactile cues. Allow children to discover for themselves whether they are right. If there are too many cubes, some will be left over; if too few, some empty spaces will be left. Guide the child in finding the right board. When there is an exact fit, the child begins to develop the concept of equality. To simplify the game in the beginning, introduce only two or three boards at a time.

2. Recognizing Number Patterns

- Give each child an empty pattern board.
- They now concern themselves with one pattern only.
- Keep the display card and cubes for your use.
- Build a cube pattern on the display card.

* Ask, "Who has the board where these cubes fit?"

- The child who thinks she or he has the board raises a hand.
- Move the display card with cubes over to that child.
- The child puts the cubes into the board.
- Children keep their boards for another game, such as the next one.

3. Building the Sequence from 1 to 6

A Calling-Back Game

- Out of cardboard, cut a 6½" x 32" strip and mark off ten sections; each section will hold a pattern board. Cut off the last four sections to use later on.
- Each child already has a board filled with cubes from the previous game.
- Keep the display card and cubes for your use.
- Build the 1-pattern on the display card.
- Ask, "Who has the pattern that looks like this?"
- Call on the child who has it.
- Ask, "Where does it live?"
- Show the child that the 1-board lives in first place on the strip.

* On your display card add one more cube to the 1-pattern, making it the 2-pattern.

- Ask the same questions for the 2-pattern.
- Continue adding one more cube to your pattern.
- Say, "Can you find the next pattern? It looks like this."
- In this way you can call all the boards "back home" without using their number names.
- Leave the boards in sequence for the next game.

4. Recognizing Like Patterns

The Dumping-Out Game

- As a result of the previous game, the boards filled with cubes are already in sequence on the cardboard strip.
- Keep the display card and cubes for your use.
- Provide a tray for the cubes to be dumped into.
- Build a pattern on the display card.

* Ask, "Who can find this pattern?"

- Call on a child to find it.
- Say, "Dump it out."
- The child puts the board, now empty, back in its place.

Note: Conservation of quantity--In dumping out the cubes of a pattern, the children see that the number of elements remains the same even though they occupy more space as they scatter. Reversibility--Ask, "Do you think the cubes you dumped onto the tray will all fit back into the same pattern board? Try it!"

5. Building Number Patterns

"Action is more conducive to the formation of images than is mere perception."--Piaget

- Explain to the children that they are going to build with cubes the pattern you show them.
- Give a child some cubes and a display card (with star, colored margin on the left, and grid for 10 cubes).

* Hold up pattern board 3.

- Say, "Build this pattern with cubes."
- Encourage them to build in this order: "left cube, right cube"; then the next row: "left cube, right cube."
- When they are finished, they check their accuracy by fitting the cubes into the board you hold.
- Give each child his or her board to keep until all six boards are filled.

Group Game (not illustrated)

- Give each child some cubes on a little paper mat and a display card.
- Hold up a pattern board; then hide it.
- Each child must build the same pattern from memory.
- When everyone has finished building her or his pattern, show the pattern board for checking.

6. Checking Up: Recognizing Patterns 1 to 6

- Scatter empty pattern boards face up on a table across the room.
- Seat the children around you at a table.
- Build a cube pattern on your display card and explain that the children are to get the board in which the pattern fits.

* Call on a child; ask, "Where will these cubes fit?"

- The children watch to see which board will be brought back.
- Have the child place the cubes in the board as a check.
- The child should empty the board and return it so as to keep a large choice of boards on the table across the room.

Note: Some children may need help. Before they cross the room, have them touch your cube pattern. This fosters concentration and builds visual memory.

Introducing Patterns 7, 8, 9, and 10 (no number names yet)

Patterns 1 to 6 are already known. Patterns 7, 8, 9, and 10 will be easy for the children to recognize. Introduce one new pattern at a time. Place it with the familiar boards, then play one of your favorite games.

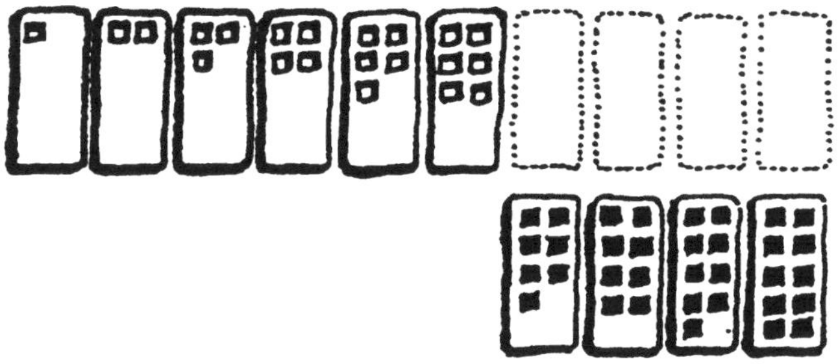

Building Patterns on the Display Card

Cube patterns 7, 8, 9, and 10, when built on the display card, become instantly recognizable. The crisscross grill of lines enables the children to see the 7-pattern as the full board with 3 cubes missing, the 8-pattern as the full board minus 2 cubes, and the 9-pattern as the full board minus 1 cube. The full board, or 10, is very easy. Present them this way. Then play the games you have done for patterns 1 to 6.

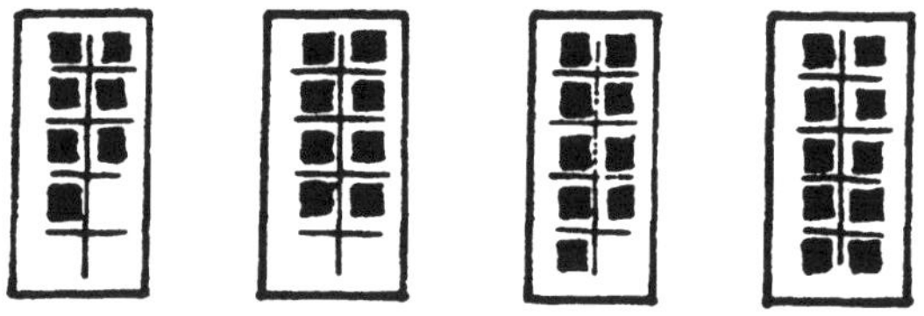

Building Patterns with Subgroups

It is also fun to build the patterns on the open table top without the boundaries of the display card to help. In this case the pattern must be built with two colors of cubes in small subgroups.

7 is 6 and 1

8 is 4 and 4

9 is 8 and 1

10 is 6 and 4

Abstracting Number from Materials

Children seem to have the innate ability to understand that the concept of a certain number can be thought of separately without being connected to the cubes.

One day the teacher can pass out the pattern boards to the children and build the familiar patterns out of other objects, such as pennies, beads, or beans. The child with that pattern board can then claim the matching bean or penny pattern. Ultimately, the number concept behind the symbol *4* or *6* has to exist as an abstract concept that can be moved about in the mind. Children need not be able to put a concept into words. The teacher can see that it exists if they can match patterns that are alike. This ability to see likenesses and differences is important as a reading readiness skill as well.

Flexible Number Groups

Children love to recognize small groups of objects dropped haphazardly on the table. They see the amount and match it to the pattern board. If they do not, the teacher can rearrange the flexible group into the structure of the pattern on the pattern board, helping to develop the number sense that each of us is born with. It can be a challenging game even before the number patterns have their names and before children can count.

7. Introducing Patterns 7 to 10

Introducing the 7-Pattern

- Lay out the cardboard strip extended to ten sections.
- Keep the display card and cubes for yourself.
- Pass out pattern boards 1 to 7.
- Build patterns 1 to 6 on your card.
- Have children fill pattern boards 1 to 6 and put them in place.
- * Then say, "Watch! Which pattern comes next?"
- Start by building the 6-pattern; say, "It's this pattern with one more cube." (Add the last cube.)
- The child who has the 7-board fills it and puts it in its place on the strip.

Introducing Patterns 8, 9, and 10 (not illustrated)

- Leave pattern boards 1 to 7 filled and in place.
- Scatter boards 8, 9, and 10 on the table.
- Build the 10-pattern on your card.
- Ask, "Who has this pattern? Where does it live?"
- A child will recognize the full board; fill it and put it last in the sequence.
- Build the 10-pattern on your card again, but remove the last cube.
- When a child recognizes it, have him or her fill the 9-board and put it in its place next to last.
- Build the 10-pattern again but this time remove the last 2 cubes. Have a child fill the 8-board and put it in place.
- Now all ten patterns have been introduced.

8. Checking Up: Recognizing Patterns 1 to 10

Stage-Setting Game

- Lay out the cardboard strip of ten sections.
- Give out pattern boards 1 to 10.
- Keep the display card and cubes for yourself.
- Build on your card an easy pattern, for example, the 4-pattern.
- The child with the 4-board puts the cubes into it and pushes it into place on the strip.
- When patterns 1 to 6 are in place, build the 10-pattern.
- * Ask, "Who has this pattern? Where does it live?"
- A child fills the 10-board and puts it in last place.
- Present the 8-pattern and explain that it looks like the 10-pattern with 2 cubes missing.
- Present the 9-pattern and explain that it looks like the 10-pattern with 1 cube missing.
- Finally, have a child put the 7-board in place.

The Sticker Game (not illustrated)

- Have the children close their eyes.
- Hide a sticker under a pattern, for example, the 9-pattern.
- Build the 9-pattern on your display card.
- Say, "Open! The sticker is under this pattern!"
- The child who looks beneath the like pattern board will find the sticker and can keep it.

The 10-Box: Level I

Introduction

The 10-Box

(See illustrations on page 17.) The length of the 10-box represents the sum total of 10; the blocks represent the numbers. By fitting blocks together in the 10-box, children discover that there are ten different combinations with a total of 10. They gain insight into the following number combinations: 1 + 9, 2 + 8, 3 + 7, 4 + 6, 5 + 5, 6 + 4, 7 + 3, 8 + 2, 9 + 1, and 10 + 0.

These facts are studied first because the base of our number system is 10.

Level I. In the beginning, the children fill the 10-box with blocks as though they were working on a puzzle. Experimenting with these materials provides the children with experiences on the sensorimotor level. They should understand these relationships before they use number names or symbols.

Experiments in the 10-Box. Present the box already filled with all the blocks so the children can see how the finished task should look. Then remove the blocks and place them on a table. Start the game by putting a biggish block, such as 8, on the left side of the box and invite the children to fill the empty gap above the 8-block (8 + 2). If they add a block that is too big (8 + 3), encourage them to look for a block that just fits. Sometimes they fill a gap with two smaller blocks. Show them how to replace these blocks with a single block. They must do this to keep the little blocks free, for they will be needed to go with other big blocks.

What Kind of Thinking Is Going On?

Many mathematicians describe mathematical thinking as often spatial and visual. This describes the thinking process of the children using these materials. Each time they put a block in the box, they measure with their eyes to judge the size of the number block that will fit with it. They systematically study the relationships between each separate number block and the total, 10. They see that when the first addend is big, the second one must be smaller, and vice versa. When they make a mistake, they can feel and see in what way that block does not fit. A teacher doesn't need to say, "Wrong." They see that the block is wrong by a certain amount and this tells them what kind of change to make. If the block is too big, that directs them to look for a smaller block. When they find a block that just fits, they feel a satisfaction that is unforgettable. This success helps them remember the combinations. They enjoy thinking for themselves. They know the discoveries they have made are reliable. They have learned to perform a task by gaining insight into its structure. Later they will be able to reconstruct any forgotten fact by recreating an image of it in their minds. They will use these images to reason mathematically.

The Number Boxes from 1 to 9

When any one of these number boxes is filled with combinations of blocks, it yields all the facts with the sum of that number box. For example, the children might select the 6-box and fill it as though it were a puzzle. Later they will record these combinations as addition facts: 5 + 1, 4 + 2, 3 + 3, 2 + 4, 1 + 5, and 6 + 0 (since the 6 block fills the entire 6-box by itself). They now have 6 combinations that make 6.

Number Names

At first, do not refer to the boxes by their number names (the 6-box, etc.). Do not give the blocks their number names (the 6-block, etc.). In carrying out these experiments, children are studying the relationships between the number blocks and the totals. It is more important for them to concentrate on relationships; they can learn the names later.

On Level II the children will seriously study the combinations of blocks in each box. Once they know the number name of each block, they will be able to recite all 55 addition facts with sums of 10 or less. On Level I let the children enjoy filling the boxes for fun.

1. Filling the 10-Box (no number names yet)

- Start with the box filled with combinations of blocks.
- Dump all the blocks onto the table.
- Hold up the empty box so that it faces the children.
- Fill the first row as a demonstration (use any two blocks, such as the 7-block and the 3-block); begin on the children's left.
- Next, put in a big block, perhaps the 8-block.

* Say, "Now, I put in a block for you. Can you find the block that fits in here?" (point).

- Call on a child. Continue filling the box in the same way.

Note: The combinations of blocks are put together as in a puzzle. Encourage each child to experiment to find the block that fits. Put a big block in first; it is easier to estimate a gap the size of 1, 2, 3, or 4 units.

2. Filling the 10-Box

Concept: *no other block*

- Start with the box filled with blocks.
- Dump out the blocks; give each child several small blocks (pass out blocks 1, 1, 2, 2, 3, 3, 4, 4, and 5).
- Keep the big blocks: 5, 6, 6, 7, 7, 8, 8, 9, 9, and 10.
- Fill the row on the children's left with the 10-block; elicit that no block can fit with this block.
- Next, put in a big block.

* Ask, "Who has the block that fits here?" (point).

- Continue putting in a big block and giving children turns completing the row.

Note: Should a child try the wrong block, wait; she or he may see that it doesn't fit and remove it. Ask, "Does it fit?" Encourage children to think for themselves.

3. Finding Like Pairs of Blocks

The Twin Game

- Display the 10-box filled with blocks.
- Dump out the blocks on the table.
- Place any combination of blocks in the first row (except the 10-block or the two 5-blocks).
- * Ask, "Who can find the twins of these blocks?"
- Help with this first demonstration set, if necessary.
- Continue putting in combinations of blocks to be matched by their twin pairs of blocks.
- Encourage children to admire the box, now filled in a different way.

Note: It brings great pleasure to children when they are allowed to discover on their own that a combination of blocks occurs twice (except for the 10-block and the two 5-blocks). Children also enjoy this game as a group game.

4. Checking Up: How Many Pairs of Blocks Are Known?

- Do this only after many games and experiments.
- Display the 10-box filled with blocks.
- Dump out the blocks.
- Put in the 10-block; elicit that it fills the row by itself.
- * Ask, "Can you find two blocks that go together?"
- Call on a child.
- Some children are able to select two blocks that go together while they are still outside the box.
- Continue until the box is filled with combinations.

Note: If some children are still unsure, they will need help with hard combinations (7 + 3; 6 + 4). Place the big block in the box first; then let a child fill in the small gap. Continue until all the combinations are in the box.

5. Learning about the Structure of a Stair (not illustrated)

- Build a stair from 1 to 10 in the 10-box.
- Give children turns climbing your stair with one finger.
- Encourage them to sing "Up, up, up" in time to climbing each step.

Repairing the Stair

- Build a stair from 1 to 10 in the 10-box.
- Scatter the rest of the blocks on the table. (You will need to add another 10-block from the counting board.)
- Tell the children to close their eyes.
- Remove one of the blocks from your stair.
- Place it among the blocks on the table.
- * Say, "Open your eyes."
- Ask, "Who can find the missing step?"
- Call on a child; let him or her feel the gap and put back the missing block.

Note: When the stair has been repaired, have the children climb it with one finger. If they can say, "Up, up, up," saying one word for each step, they are ready to learn to count things.

6. Building the Stair (Forming a Sequence)

- At first, use only one set of blocks 1 to 10. (Later use all 19 blocks.)
- Scatter blocks 1 to 10 on the table.
- Hold up the empty 10-box.
- Begin the stair in the lower left corner with blocks 1, 2, and 3.
- * Ask, "Which block comes next?"
- Say, "Is it right? Climb up with your finger."
- Give children turns putting in the next step of the stair until it is finished.

Note: This is a difficult task. There is no groove to indicate the size block being looked for. Once children realize that the next larger block is at the same time the smallest in the group of remaining blocks, they have the key to the situation. Encourage them to experiment. They often try a block the same size as the previous step, then one too big, then too small, and finally the right one. Let them test the stair by climbing up with one finger. At some time, let each child build the stair alone.

7. For the Child Who Has Difficulty Building the Stair

- Arrange a set of blocks from 1 to 10 on the table so that the end of each block is aligned with the table edge.
- Build the first steps vertically in the box with blocks 1, 2, and 3.

* Say, "Look at all these blocks."
- "Pick the smallest block."
- "Put it in the stair."
- Children can see this works; they gain the confidence needed so that they can build the stair by themselves.

Note: Some children cannot construct a stair that rises evenly one step at a time. This experiment gives them the kind of learning experience that helps them comprehend the structure of a stair. They experience each time "the one longer than" and find they are building the stair by themselves.

8. Checking Up: Building the Stair

A Group Game (child directed)
- Scatter two sets of blocks 1 to 9 and the 10-block on the table.
- Place the 10-box in front of the first child.
- Start the stair with blocks 1, 2, and 3.

* Say, "Put in the next step. Pass the box to your neighbor. Ask him . . . . ?" (Whisper, "What comes next?")
- The child asks, "What comes next?"
- The box is passed from child to child until the stair is completed.

Filling in the Stair (not illustrated)
- Turn the box with the completed stair so that the space for the "inverse stair" is at the bottom of the box.
- Show the children how another stair can be built by putting the 1-block with the 9-block, and so on. Children continue until the 10-box is completely filled with blocks.

Checking Up: Level I

Exploring Relationships and Making Discoveries

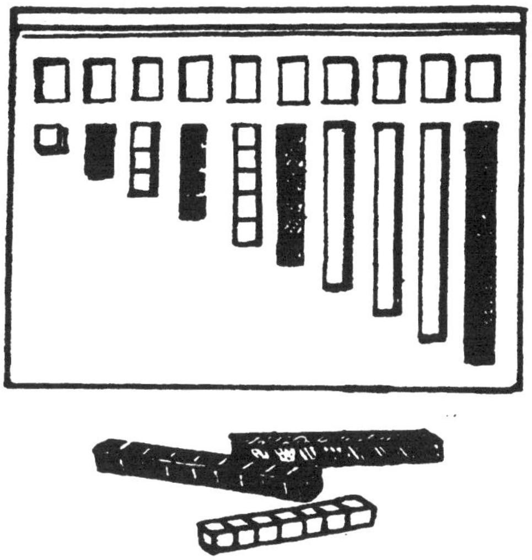

The children's number sense and mathematical reasoning have been developed through working with these materials as though they were puzzles. We see this in their awareness of small and big and in their ability to recognize the role that a block plays in the building up of a total. Before number names and numerals are presented, the foundation for the later comprehension of arithmetic has been laid down. On Level I the children should be able to perform the following tasks:

The Counting Board (no number names yet)
- To put each block in the groove where it fits
- To find the block for the groove you indicate
- To realize when a block is too small or too big for the groove and find the one that fits
- To know where each block "lives" in the sequence

The Pattern Boards and Cubes (no number names yet)
- To recognize the matching pattern board for any cube pattern you build
- To be able to build the pattern with cubes for any board you show them
- To build the sequence from the smallest to the largest amount
- To know where each pattern "lives" in the sequence

The 10-Box (no number names yet)
- To fit together the blocks so they form the combinations that fill the 10-box
- To build a stair with the blocks from the smallest to the largest
- To find the missing block when one step has been removed.

The Counting Board: Level II

Introducing Number Names, Counting, and Language

Goals

The children should be eager and ready now to talk about what they have discovered in their experiments. To do this, they need to know the name of each number block and pattern; they must learn to count. On Level II they learn to do the following tasks:

- To count from one to ten
- To name each block, even when seen outside the counting board
- To know which number block is meant when described with these words: *one bigger than, one smaller than, after, before, between,* and *equal to*

Counting

In the usual teaching of counting, the teacher presents children with the task of counting like elements in a scattered group. This piecemeal approach does not lead to the children's seeing relationships between the numbers. However, counting does enable one to know the number of items in a group and is certainly a skill needed in daily life. In Structural Arithmetic, counting is not used in computation. It is used to figure out the names of the blocks and the position of a numeral in a series.

Learning to Count: Finding the Name of a Block. An analysis of the process of counting will help teachers assist those children who find it difficult. The prerequisite for counting is mastery of the sequence of number names. Even when they know the sequence of names, some children still have trouble counting objects. Those who fail to recite one name when pointing to one object need to learn the correct timing.

You can control the factor of timing by reciting the words with the children. As you fill a groove of the counting board with cubes, join the children in reciting the counting words. Next, teach children to enumerate cubes in time to the tapping of a stick. Once they can count, let them discover the name of each number block.

Learning to Count by Naming the Blocks in the Counting Board. For children who continue to find learning to count difficult, provide more help in mastering one-to-one correspondence. One way of teaching them to count is to have them name each block in succession in the counting board. Using one finger, they climb up the block stair; they touch the 1-block and say "one," touch the 2-block and say "two," touch the 3-block and say "three." Each block acts as a monitor that helps these children keep together the naming of the block with the touching of the block. The teacher might remind a child, "You are touching the 3-block and calling it 'four.' Let's start again." As a result, children learn to coordinate the action of pointing with the saying of the number name.

Avoiding the Use of Color Names

Do not refer to a block by its color. In this situation color is only one aspect of the block. We want the children to be more aware of the characteristics that are important, that is, the *size* of the block and its *place* in the sequence. If they mention the block's color and you do *not* use this name, they will not use color names nor will they feel rebuked.

Cardinal and Ordinal Aspects of Numbers

As children count by naming each block in succession, they are at the same time seeing the size of the number name they are saying. The size is the cardinal aspect of the number. They also discover, for example, that the 4-block lives in fourth place; this is the ordinal aspect.

Seeing Relationships

Children who have been taught to count as the only way of dealing with addition learn to do nothing but count. Counting dissipates their focus; they don't see relationships between the numbers.

Working in the counting board is a more enriching experience. Children "meet" each number block as part of an organized system, the number sequence from 1 to 10. If they count in the "up" direction, they learn that each block is one bigger than the preceding block. If they count in the "down" direction, each number block is one smaller than the block just named. At the same time, they learn the name of the block that comes *before* a certain block and then the name of the block *after* a block. These relationships are basic to understanding the mathematical concepts of adding 1 to, or subtracting 1 from a number.

Language and Experience

One of the most important ways of having children learn more about a concept is to put them into the situation of being teacher. In this way they learn to put their ideas into their own words. For example, there are several ways to ask for a 6-block and a 5-block. Will they say, "Get 5 and the block one bigger than 5?" Or will they say, "Get 6 and the block one smaller than 6?" Each idea helps them understand the opposite idea better; each reinforces the other. Using language helps children deepen their concepts.

Children have only partially grasped their experiences with the materials on the manipulative or sensorimotor level. The language children hear on Level II and the words they find themselves using will help them further develop these concepts.

1. Learning to Count--Discovering the Names of the Blocks

- Remove the blocks from the counting board.
- Fill the 1-, 2-, 3-, and 4-grooves with cubes of one color for each row.
- Each row of cubes is called a mock block. The mock 4-block might be all red cubes.
- Point to the mock 4-block.
- * Say, "Find a block the same length as this row of cubes."
- A child places a block on top of the row of cubes and sees that they are equal.
- Say, "Now count out the cubes."
- Guide the child in counting out the cubes.
- The child removes a cube as she or he says each number name, "One, two, three, four."
- The child replaces the 4 cubes with the 4-block.
- The children have discovered that this block is called "four."
- Have them discover the name of each block in the same way.

Note: Some children get so attached to the act of counting objects that they can't stop doing it. They repeatedly count the units in blocks they know the names of. Teach these children by having them name each block in succession in the counting board (see next experiment).

2. Learning to Count--Identifying Blocks by Their Number Names

Review: Display the counting board filled with blocks. Review counting by saying the names of the blocks. Point to the blocks in succession as children say, "1, 2, 3, . . . 10." This controls the one-to-one correspondence between one item and one number name.

"Stand Up —" (name a block)

- Name the blocks in order as you call on children to stand them on end in the counting board.
- * Say, "Stand up 4," and so forth to the last block.
- Later on, you may name the blocks out of order.

"Put Down —" (name a block)

- Name the large blocks first before they tip over.
- Call on a child to put each block down.
- Continue until they are all down.

Note: When children cannot identify the blocks you are naming, help them. Show them they must start with the 1-block and name each block in order until they hear themselves say the name you are asking for. If the teacher says, "Find 5," the child responds, "1, 2, 3, 4, 5!"

3. Selecting a Named Block

A Calling-Back Game (Easy Version)

- Display the counting board.
- Give each child several blocks.
- Point to the empty 5-groove.
- * Say, "Calling back 5!"
- The child who has the 5-block returns it to its groove.
- Continue until each block is in its place.

A Calling-Back Game (Difficult Version) (not illustrated)

- Remove the blocks and scatter them on the table.
- Do not point; use oral cues only.
- Ask, "Who can find the 5-block?"
- The child who finds the 5-block returns it to its groove.
- Continue until they are all in place.

Note: This is not a test. Encourage self-teaching. If the children do not know the block named, such as "5," have them count the places in the board. They will hear themselves say "5" and see its size.

4. Do You Know the Name of This Block?

- Display the counting board filled with blocks.
- Give each child a block, reviewing the name of the block as you give it out.
- Point to the empty 10-groove.
- * Ask, "Who has the block that lives here?"
- The pupil who has it returns it.
- Ask, "Do you know its name?"
- Encourage children to discover the name by themselves (counting spaces in the board).

The Hiding Game (not illustrated)

- Display the counting board filled with blocks.
- Say, "Close your eyes"; then hide the 10-block behind your back.
- Say, "Open your eyes. Which block is missing?"
- A child says, "10."
- Give the child the hidden block so he or she can replace it in the board.
- Let the children take turns being teacher.

5. Learning about "One Bigger Than"

Teaching the Relationship (not illustrated)

- Display the filled counting board.
- Say, "Show me the 1-block."
- Say, "Stand up the block one bigger than the 1-block."
- A child stands up the 2-block; let him or her feel the one step up from the 1-block to the 2-block.
- Say, "Please point to the 2-block. Stand up the block one bigger than 2."
- A child stands up the 3-block and then climbs up the one step with his or her finger.
- Continue until all the blocks are standing up.

Practicing the Relationship

- Leave the blocks standing up.
- Tell the children to listen to your directions; you will tell them which block to put down.
- Call on a child.

* Say, "Point to 4. Put down the block one bigger than 4."

Note: If children are having difficulty, let them feel how much "one bigger" is while the two blocks are standing side by side.

- Ask, "What is its name?" (Five.)

6. Learning about "One Smaller Than"

Teaching the Relationship (not illustrated)

- Display the empty counting board.
- Place the blocks on the table.
- Say, "Put back the 10-block."
- "Now, put back the block one smaller than the 10-block."
- Let the children feel one step down with a finger.
- Continue until all the blocks are in place.

Asking for Two Blocks, Stating Their Relationship

- Place the blocks in random order on a table across the room.

* Say, "Listen hard! Get 5 and the block one smaller than 5."

- Soon all the blocks will be back.

7. Learning about *After*

- Display the counting board.
- Scatter the blocks on the table.

* Call on a child.

- Say, "Point to 4. Can you put back the block that comes after 4?"
- Ask, "What is its name?"
- A child answers, "5."

Note: When a child is stuck, point to the empty groove to show where the position "after 4" is. Once the children understand the meaning of *after,* play the game without pointing to a groove.

8. Learning about *Before*

Note: For some children the concept of *before* is very difficult. In this experiment the meaning is put into the context of time: "What went before this?"

- Display the filled counting board.

* Place a 10-box near the beginning (left side) of the board.

- Announce, "The blocks are going into this box. Watch! First the 1-block goes into the box."
- Place the 1-block in the box as the first step of a stair.
- Say, "Now 2 goes into the box."
- Ask, "Which block went before 2?"
- A child answers, "One."
- Ask, "What am I going to do next?"
- A child says, "Take the 3-block" (teacher does so).
- Encourage the children to tell you what to do. You just move the blocks.
- Each time ask, "What went before . . . ?"
- Soon the blocks are in the 10-box in the form of a stair from 1 to 10.

9. Learning about *Between*

Note: Some children have difficulty understanding where the position *between* is. It may be easier for them to understand when you ask them, "What is between your hands?" Try it.

Teaching the Relationship

* Guide a child by saying, "Put this hand on 4. Put this hand on 6."
- Ask, "What comes between 4 and 6?" (or "What comes between your hands?").
- The child looks at what is between his or her hands, sees the 5-block and names it.

Practicing the Relationship (not illustrated)

- Give both points of reference by saying, "Put back 4. Put back 6."
- Then say, "Put back the block that lives between 4 and 6."
- Ask the child to name it. (Five.)
- Continue with other sets of blocks.

10. Checking Up: The Sticker Game

- Buy small stickers for the children to win.
- Display the filled counting board.
- Say, "Close your eyes."
- Hide a sticker underneath the 4-block.
- Have the children open their eyes.
* Say, "The sticker is under the block *before* 5."
- Ask, "Which block comes before 5?"
- The child locates 5, finds the 4-block is before 5, looks underneath it, and finds the sticker.

Checking up: You may vary the vocabulary; use *before, after, one bigger than, one smaller than,* and *between.*

Pattern Boards: Level II

Introduction

Goals

- To name each number pattern
- To understand the characteristics of odd and even numbers
- To name the number that results when any of the following actions are demonstrated with cubes:

  Adding 1 to a number
  Subtracting 1 from a number
  Adding 2 to a number
  Subtracting 2 from a number
  Adding or subtracting 0 from a number

Level I: What Was Learned?

Children seem to have good number sense. They can recognize a pattern by its characteristics and can learn to reproduce it with cubes without counting. They developed this number sense on Level I by matching and building number patterns.

Level II: Giving Patterns Their Number Names

The number patterns show children another way to visualize numbers. The patterns bring out different characteristics than do the number blocks. For example, the 4-block is just a block one unit shorter than the 5-block. In contrast, the 4-pattern looks square and quite different from the 5-pattern, which is an odd number and therefore has an extra cube. Since their characteristics are more pronounced, the patterns are easier to remember than the number blocks. Children quickly recognize and name patterns 1 to 6; the 10- and 9-patterns, being at the end of the sequence, are readily learned. Counting should be discouraged. Recognition of a pattern is learned through building it with subgroups of different colored cubes, such as 4 and 4, or comparing its size with the full 10-pattern.

Operations and Actions

Concepts developed through experiments in manipulating the materials have probably been only partially grasped by the children. The language they will now hear and the words they will use themselves will help them master these concepts slowly over a period of time.

Piaget believed that mathematical concepts do not arise from the materials themselves but from an appreciation of the operations that can be performed with the materials. Children acquire mathematical concepts from working with the Structural Arithmetic number patterns and soon gain the ability to move the concepts around in their minds.

Structurally Related "Families" of Facts

Some groups (or "families") of facts belong together because they have certain characteristics in common.

Adding 0 to a Number. It is easy to show children that if you add nothing (or no cube) to a pattern, it stays the same.

Adding 1 to a Number. Adding 1 to a number yields the next higher number. Children are amazed that adding one more cube to a number pattern causes the next higher pattern to appear. At the same time, they see an even number become odd or an odd number become even by the addition of one cube. Once they can put into words what adding 1 to a number does, they have laid the foundation for the mastery of these facts.

Subtracting 1 from a Number. Subtracting 1 from a number yields the next lower number. It is important to understand the reverse of a concept as it is a basic aspect of mathematical thinking.

Children should know the facts that result from adding 1 to a number or subtracting 1 from a number and be able to recite them.

Odd and Even Number Patterns

The quality of evenness or oddness cannot be shown with the number blocks. The pattern boards are specially designed to show the characteristics of odd and even numbers.

Even patterns are studied first. All children seem to know the doubles 1 + 1, 2 + 2, 3 + 3, 4 + 4, and 5 + 5. *Odd patterns* are presented as related to even patterns. Most children see the 5-pattern as 4 + 1 and say, "That pattern has an extra cube."

Most workbooks show other patterns, but they do not demonstrate the desired relationships. For instance, dominoes are inadequate: domino 5 seems so self-contained it is difficult to see how adding 1 dot to it would produce the domino 6-pattern. Domino 7 looks so different from domino 8 that the difference of 1 is not readily perceived.

Adding 2 to an even number yields the next higher even number.
Adding 2 to an odd number yields the next higher odd number.

On Level II, children act out these statements and put their discoveries into words. This oral work helps them develop the concepts that will enable them to read and understand the written examples to be studied on Level III.

Subtracting 2 from an even number yields the next lower even number.
Subtracting 2 from an odd number yields the next lower odd number.

Subtracting is the inverse operation of adding. On Level II, children act out the statements given above and put their discoveries into words.

1. Naming the Pattern Boards and Putting Them in Sequence

- Lay out the cardboard strip with ten sections.
- Give the pattern boards to the children.
- Keep the display cards and cubes.
- Begin with 1 cube on the display card.
- The child with the 1-board names it, puts a cube into it, and pushes it into first place.

* Build the 2-pattern on your display card.

- Ask, "Who has this pattern board? What do you think its name is? Where does it live?"
- A child says, "2," puts the cubes into the 2-board, and pushes it into second place.
- Continue until all ten pattern boards are named correctly and put in place.

Note: Children have learned the number names and how to count in the experiments with the counting board. They should be able to recognize the number patterns without counting. Allow children to verify them by counting, but then discourage more counting. Encourage them to see each pattern as a whole.

This game can also be played in random order.

2. Recognizing and Building Patterns

Note: "Action is more conducive to the formation of images than is mere perception."--Piaget

- Seat the children around a table.
- Give each child a display card and cubes on a round paper mat.
- Advise the children to watch you carefully because you will ask someone to be the leader after you.
- Build a number pattern on your card.
- Ask all the children to copy it.
- Check each child's pattern.
- Now select a leader.

* The leader builds a pattern such as 5.

- He or she says, "Build this pattern. Build 5."
- All the children build that pattern.
- The leader checks each pattern.

Note: The leader may not recognize from across the table that a pattern is correct. To help, turn the unrecognized pattern so it faces the leader. It is important that children learn to recognize the patterns from every angle.

3. Adding 1 to a Number Yields the Next Higher Number
Adding 0 to a Number Results in That Number

Vocabulary: *zero*

- Lay out the cardboard strip with ten sections.
- Give out pattern boards to the children.
- Keep the display card and cubes for yourself.
- Show the empty display card.
- Say, "There's nothing here; we call that *zero*. I add 1 cube.
- What do you see?" ("1 cube.")
- Have the child with the 1-board fill it and put it in place.
- * Put 1 cube on your display card.
- Say, "I have 1; I add one more cube. Who has 1 and 1? What is its name?"
- A child names it, "2," fills the 2-board, and puts it in place.
- Put 2 cubes on your display card.
- Say, "I have 2; I add one more cube."
- Ask, "Who has 2 and one more? What is its name?" ("3.")
- Continue until you reach "9 and one more is 10."

Note: Another new idea learned in this experiment is that each number can be called by "another name." The teacher names the 2-pattern "1 and 1" and calls the 3-pattern "2 and 1," and so forth. It is important to contrast adding 1 with adding 0 (zero or no cubes) to a pattern; in the latter case, the pattern remains the same.

4. Subtracting 1 Yields the Next Lower Number
Subtracting Any Number from Itself Leaves No Remainder

- Lay out the cardboard strip with ten sections.
- Give out pattern boards to the children.
- Keep the display card and cubes for yourself.
- Fill the 10-board with cubes.
- Have a child place it at the end of the sequence.
- * Build another 10-pattern on your display card.
- Say, "Here is the 10-pattern. Watch! I take 1 cube away. What is it now?"
- The child with the 9-board names it, "9," fills it with the cubes, and pushes it into the place before the 10-board.
- Continue with 9 minus 1, 8 minus 1, . . . 2 minus 1.
- Introduce 1 minus 1, which leaves no cubes at all, or *zero*.

Note: This experiment lays the foundation for children's understanding later that subtraction is the inverse of addition. Here children see that what adding 1 "does up," taking 1 away "undoes."

5. Learning about Even Number Patterns

Note: Before playing this game, teach the concept of partners. For instance, have each child take a partner when standing in line.

Teaching the Concept of Even Numbers

- Have all the boards filled with cubes and put in place by playing game 3 or 4, page 30.
- * Move the first even board (2) down.
- Say, "These 2 cubes are partners. We call this pattern even.
- Find another even pattern!"
- Call on children to move the even patterns down, one after another.
- Remove the odd pattern boards (or cover each odd board with an index card).
- Have children name the even numbers and leave the boards in place for the next game.

Finding Even Number Patterns (not illustrated)

- Play the Dumping Out game, as follows.
- Call on children to name and dump out the biggest even number (10) and the smallest even number (2).
- Ask, "What's the biggest even number now?" (8) etc.

6. Learning about Odd Number Patterns

Teaching the Concept of Odd Numbers

- Have the pattern boards filled with cubes and put in sequence by playing game 3 or 4.
- * Point to the 3-board and move it down (the 1-board is too difficult to start with).
- Explain, "These 2 cubes are partners. This cube has no partner" (point to it).
- "We call this pattern odd."
- Say, "Find another pattern with 1 cube all alone."
- A child finds it, names it, and moves it down.
- Continue until all the odd patterns have been moved down.
- Go back and move down the 1-board.
- Remove the even boards, leaving a sequence of odd number patterns (or cover each even board with a card).
- Help the children see that every other number is odd.

Finding Odd Number Patterns (not illustrated)

- Play the Dumping Out game, as follows.
- Call on children to name and dump out the biggest odd number (9), and the smallest odd number (1).
- Ask "What is the smallest odd number now?" (3) etc.

7. Adding 2 to an Even Number Yields the Next Higher Even Number

- Lay out the cardboard strip with ten sections.
- Call on children to put in place each even pattern board.
- Remove the odd pattern boards.
- Give a display card and cubes to each child.
- * Say, "Let's all start with the 2-pattern."
- Now say, "Watch! What am I doing?"
- Add 2 more cubes to your 2-pattern.
- Say, "Now, you do it!"
- Children place 2 more cubes on their cards.
- Ask, "What did you do?"
- A child answers, "I added 2 cubes."
- Ask, "What is it now?"
- A child answers, "4," puts the cubes into the 4-board, and is given 4 new cubes.
- Next, add 2 cubes to your 4-pattern.
- Continue until all the even boards have been filled with cubes.

Note: The teacher elicits the action of adding 2 cubes and the children express the action in their own words. This helps them understand the operation of addition. Play the same game with the odd numbers and with subtracting 2 cubes. This activity is worth the effort.

8. Subtracting 2 from an Even Number Yields the Next Lower Even Number

- Lay out the cardboard strip with ten sections.
- Pass out the even pattern boards.
- Keep the display card and cubes for yourself.
- Build the 10-pattern on your card.
- The child who has the 10-board names it and puts it in last position.
- * Build the 10-pattern again.
- Say, "Here is the 10-pattern."
- Remove the last 2 cubes.
- Call on the child with that board.
- Ask, "What did I do?"
- The child answers, "You took 2 cubes away."
- Ask, "What is it now?"
- The child answers, "8!" and puts the 8-board in place.
- Continue until all even boards are in place.
- End by taking 2 cubes from 2 cubes; elicit that no cubes are left.

Note: This helps children understand that in subtraction we start with a certain amount; then a number is removed, and a new amount remains behind (later called the remainder).

9. Checking Up: The Sticker Game

Vocabulary: contrasting *one more than* with *one less than* and *one bigger than* with *one smaller than*

- Buy small stickers for the children to win.
- Have children put the empty pattern boards in sequence.
- Keep the cubes, display card, and stickers for yourself.
- Tell the children that when their eyes are closed, you will hide a sticker beneath a board.
- Say, "Close your eyes." Hide a sticker under the 6-board.

* Build the 7-pattern on your display card.

- Say, "Open your eyes. Listen hard!
- I hid a sticker under the board one smaller than the 7-pattern."
- A child says, "6," lifts up the 6-board, and claims the sticker.
- Continue as long as the game is fun.

Note: Practice one set of contrasts at a time. Contrast adding 1 with subtracting 1. At another time contrast adding 2 with subtracting 2. If the children don't understand the description "one more than," have them add 1 cube to your pattern. And likewise, if "one less than" is difficult, have them take away 1 cube from your pattern so that the pattern they want will emerge.

10. Checking Up: The Screen Game

- Use an empty 10-box or a piece of cardboard as a screen.
- Keep the cubes and display card for yourself.
- Seat the children so they cannot see in back of the screen.

* On the display card build the 6-pattern with cubes.

- Ask, "What is the name of this pattern?" ("6.")
- When the 6-pattern is behind the screen, remove the last 2 cubes.
- Reveal the changed pattern on the other side of the screen.
- Ask, "What did I do? What is left?"
- A child answers, "You took 2 cubes away; it's 4."
- Continue building new patterns and subtracting 1 or 2 cubes or adding 1 or 2 cubes.
- Try taking away all the cubes to show the concept of zero as remainder.

Note: Sometimes children cannot remember how the original pattern looked before the teacher changed it. If that is the case, leave the original 6-pattern untouched. Make a new pattern so they can compare it with the original pattern.

The 10-Box: Level II

Introducing Number Games with Number Names

Goals

- To name the combinations of blocks that go together to fill the 10-box and express them orally as equations. We call this "Telling the Story of 10."
- To understand that the order of the block-addends makes no difference to the sum (8 and 2 or 2 and 8)
- To understand the concept and language of the missing addend (8 and what makes 10?)
- To fill the number boxes from 1 to 9 with combinations of blocks and to tell their "stories," but not from memory

Naming the Block Combinations

The experiments on Level I made children familiar with the pairs of blocks ("friends") that fit together in the 10-box. As soon as they learn the name of each block, children find it easy to name the block partners: "5 and 5" or "9 needs 1." Soon they can name all the combinations of blocks, even when they have been hidden from view. Each block's own color helps children see the structure of the combinations that make 10. If all the blocks were the same color, the 10-block would not stand out, nor the two 5-blocks be detected as the double, nor the wonderful insight into the changing relationships between the addends be seen: as one addend increases, the other addend decreases. A visual analysis of the combinations that make 10 begins to take place. When a child was asked how she figured out the names of the missing pair of addends from the 10-box, she pointed to the 5 and 5 and said, "I figured it out from these." She was using her ability to reason in the best possible way.

Telling the Story of 10

Children who recite the facts mechanically often show that they don't understand the concept of equality that lies behind the facts. A child who said glibly, "Eight 'n' two're ten," was equally content when expressing the combinations before him in the 6-box as "Four 'n' two're ten." The words *'re ten* must have had no meaning but had sounded right. When children hear other children expressing these combinations, they may mimic their words without understanding what they mean. To counteract this, do everything possible to let each child express his or her discoveries using the wording that comes naturally from looking at the situation. You will know they understand the concept of equality when they say, "8 and 2 are just as big as the 10-block" while they measure them side by side. You can take the wording as proof of their insight. They will soon shorten it to "8 and 2 are 10."

Naming the Missing Addend

Playing the Hiding Game and placing the addends one in each hand makes clear to the children that if they know the sum (the length of the 10-box) and one addend, they can figure out the missing addend. The language of this game is difficult to formulate for the child who is the teacher. It is an important step in their cognitive growth as well as in their language development. A child starts by measuring the total of two addends in the 10-box and then hiding them. She or he states the total first, "I have 10 altogether." Then she or he shows one addend and says, "In one hand I have 8." Help should be given to the child to formulate the necessary question, "What is in my other hand?" This is sometimes not the way children think. They are more apt to ask, "What goes in here to reach 10?" This phrasing shows they are thinking in a mathematical way.

The zero-fact is most dramatically shown in this game. The child who hides the 10-block knows this will be different. Children often laugh and refer to it as "the joke one." The child shows the 10-block and asks, "What is in my other hand?" The natural answer is "nothing," which children understand to mean "no block." You may teach them to say "zero," which is also the name for the symbol *0*, which stands for no block and will be taught on Level III.

Telling the Story of 6

Once the children realize that the block pairs in the 10-box equal the sum 10, the length of the biggest block, they are ready to tell the story of another number box, such as 6. In looking at the filled 6-box they can spot that 6 also has a zero combination ($6 + 0$) and a double ($3 + 3$). In telling the story of 6 children often decide they don't want the same number combination to come twice and so reverse the second pair naturally: $4 + 2$, then $2 + 4$; $5 + 1$, then $1 + 5$. This shows concretely the commutative property of addition: the order of addends may be changed without changing the sum.

You will find that the language of mathematics is used naturally by children since it fits the facts they have discovered with the Structural Arithmetic materials.

1. Naming the Addends That Go Together in the 10-Box

Concept: *no block,* or *zero*

Note: The children have already discovered which blocks fit together in the 10-box. They have learned the names of the blocks in the counting board (review if necessary). You will now name one block and expect the children to find and name the missing addend. Don't teach the language of the completed equation until they understand the concept.

- ■ Empty out the blocks from the 10-box.
- * As a demonstration, put in the first combination, 7 and 3.
- ■ Say, "7 needs 3."
- ■ Put in a big block, such as 8.
- ■ Ask, "8 needs . . . ?" (point to the gap).
- ■ A child finds the 2-block, says "2," and puts it with the 8-block.
- ■ When putting in the 10-block ask, "Can any block fit with 10 in this box?"

Note: This is a unique combination. The answer is no. Children understand that no block fits with 10 in this box. Later they can say 10 and 0 (zero).

2. Naming the Addends That Go Together

- ■ Empty the blocks out of the 10-box.
- ■ Give several small blocks to each child.
- ■ Keep the 5-block and bigger blocks for yourself.
- * Put a big block into the 10-box.
- ■ Ask, "Who has the block that fits with 9?"
- ■ Say, "Good! Can you name both blocks together?"
- ■ A child answers, "9 and 1."
- ■ Continue in this way until the box is filled.

Note: Encourage the children to state each combination in their own language because this shows you that they understand it. If they say, "10 needs nothing at all" or "Ten has no friend in this box," they will be sharing their discovery with you. The words *ten* and *zero* are also the names of the number symbols children will write later on (10 and 0). It is very important for them to know the meaning of the words first before they learn the symbols.

3. Meshing Together Two Stairs

- Empty the blocks out of the 10-box.
- Ask the children to build the stair from 1 to 10 by standing the blocks up vertically.
- Guide the building of the stair from 1 to 9 in the opposite direction (see illustration).
* Carefully push down the 9-block in place.
- Ask, "9 needs . . . ?"
- A child points to the 1-block and names it, "1."
- Carefully push down the 8-block.
- Ask, "8 needs . . . ?"
- The next child pushes down the 2-block and says, "2."
- Then ask, "7 needs . . . ?"
- A child pushes down the 3-block and says, "3."
- Continue until "Nothing needs . . . ?"
- And a child responds, "The whole 10-block."

Note: Your stairway may begin with the 10-block. Push down the 10-block and say, "10 fills the box by itself." Then go on to say, "But 9 needs . . . ?" and so on.

4. Finding Two Blocks That Go Together in the 10-Box--Discovering the Commutative Property

- Dump out the blocks from the 10-box.
* Say, "Put two blocks that go together in this box."
- A child gets two "friends" and says, "5 and 5."
- Call on one child after another until the box is filled with combinations.

Note: If children select two blocks that don't go together, let them experiment. Guide them in placing the bigger block on the bottom and let them hunt for the smaller block to go with it. Letting them learn by themselves is far better than telling them the answer.

A commutative property can be discovered by the children on their own. They put in 8 and 2, then next to it, 2 and 8. This shows they understand that the order of the addends will not change the sum.

5. Overall View of the Combinations That Make 10

- Scatter the blocks on the table.
- Have the children build the stair from 1 to 10.
- They complete each row: 1 + 9, 2 + 8, 3 + 7, 4 + 6, 5 + 5, 6 + 4, 7 + 3, 8 + 2, 9 + 1, and 10 + 0.
- * Ask the children to close their eyes.
- Hide one pair of blocks (5 + 5).
- Say, "Open your eyes!"
- Ask, "Which blocks are missing?"
- A child names the missing combination, "5 and 5."
- Give the blocks to the child, who returns them to the 10-box.
- Encourage a child to be the teacher.

Note: If a child should be stuck, ask the other children what strategies they use to identify the missing blocks. They may say, "I can see the other pair of blocks in the box." You might suggest that the child count up the lower stair of blocks: "1, 2, 3, 4, 5; oh, 5 and 5!" This helps them focus attention on the different sets of addends.

6. Naming the Blocks That Go Together in the 10-Box

- Scatter the blocks on the table.
- Have the children put the pairs of blocks in the box in sequence, 1 + 9, 2 + 8, 3 + 7, . . . 10 + 0.
- Cover all the combinations with a notebook.
- Ask, "What comes first; 1 and what?"
- When the blocks have been named, move the notebook so they can check that combination.
- * Ask, "What comes next?"
- A child answers, "2 and 8."
- Reveal the blocks as a check.
- Continue until each combination has been named.

7. Telling the Story of 10

- Guide the children in expressing an equation (telling the story).
- * Place two blocks in the 10-box, such as 7 and 3.
- Measure them with the 10-block.
- As you do so, say, "Watch--7 needs 3 to be as big as 10."
- Remove the 10-block.
- Ask, "Can you put in two blocks that make 10 and tell the story?"
- A child puts in two blocks, places the 10-block beside them, and tells the story.
- Call on children to put in a pair of blocks and state the equation.

Note: Listen to what the children say. You can judge by the way they use their own words whether they understand what they are doing. Encourage them to use different expressions such as *make, are, are as big as,* and *are the same size as*. Once this concept is clear, introduce the word *equal*. The mechanical recitation of "eight 'n' two're ten," does not help children understand what an equation really means.

8. Test: Telling the Story of 10

The Scarf Game

- Dump out the blocks.
- Hide all the blocks under a scarf.
- Display the empty 10-box.
- Explain to the children that they are to ask you for blocks that make 10.
- * Say, "All the blocks are hidden. Ask me for two blocks that make 10."
- A child says, "Give me 7 and 3."
- Hand the child the blocks and ask her or him to place them in the box and tell the story, "7 and 3 make 10."
- Ask, "Is there a block that makes 10 all by itself?" ("Yes, the 10-block.")
- This is an oral test of the combinations that make 10.

Note: Sometimes it helps to allow a child who is stuck to feel the blocks beneath the scarf or to have a quick peek at part of one of the blocks.

9. Naming the Missing Addend

The Hiding Game

- Scatter the blocks on the table.
- Display the empty 10-box.
- Say, "Close your eyes."
- Now hide behind you a combination of blocks that make 10, one block in each hand.

* Say, "Open your eyes. I have 10 altogether."

- Continue, "In one hand I have 8" (place it in the 10-box).
- Ask, "What is in the other hand?" or "What does 8 need to make 10?"
- When the child answers, "2," show the block.
- Have the child complete the row in the box.

Note: Children like to be teacher in this game. Insist that they measure their two blocks in the box first as a check; the blocks must make 10. Then they may hide them and say, "Open your eyes." The most unforgettable combination is: "In one hand I have ten; what is in the other?" The natural answer is "Nothing" but the word *zero* can also be used.

10. Telling the Story of 6

- Take out the pyramid of number boxes.
- Select the 6-box.
- Scatter the blocks from the 10-box.
- Remove the blocks that won't be needed to fill the 6-box (10, 9, 9, 8, 8, 7, 7, and 6).
- Display the 6-box and put in the 6-block.
- Say, "6 needs no other block to make 6."

* Ask, "Can you find two blocks that make 6 and tell the story?"

- A child replies, "5 needs 1 to make six."
- Call on children until the box is filled.

Note: After the 6-box has been filled, dump out the blocks and play another game from this section. At this point, these boxes can be filled for fun. The children should be able to fill any box with ease. Insight into the structure of the task in the 10-box allows this transference to the other boxes.

11. Filling Number Boxes 1 to 5

- From the pyramid of number boxes, remove boxes 1 to 5.
- Give the 1-box to a child; ask what will fill it as you scatter the blocks around.
- Explain: "Yes, 1 and nothing else make 1" or "1 and zero make 1."
- Give the 2-box to a child.
- The child puts in a 2-block and then a 1-block.
- Ask, "What does 1 need to be as big as 2?"
- The child answers, "Another 1" (puts a 1-block in the box).
- Ask, "Can you tell us the story of 2?"
- The child says, "2 and nothing make 2; 1 and 1 make 2."
- Dump the blocks out of the 2-box.

* Give the 3-box to a child.

- Say, "Now, fill the 3-box."
- The child puts in the 3-block on the left, and next to it the 2-block.
- Ask, "What does 2 need to make 3?"
- A child puts in a 1-block.
- The child puts in a 1-block.
- Children fill the 4-box and the 5-box and tell the story of 4 and the story of 5.
- Play any of the other games, such as The Hiding Game or Which Blocks Are Missing.

12. The 9-Box, 8-Box, and 7-Box

Note: At first, let the children fill these boxes as though they were puzzles. The following experiments are more organized.

The 9-Box

- Present the 10-box filled with combinations 1 + 9 . . . 10 + 0.
- Place the 9-box beside the 10-box.
- Explain that you are going to fill the 9-box with the same blocks that are in the 10-box.

* Ask, "Is there a block in the 10-box that won't fit in the 9-box?"

- A child answers, "The 10-block."
- Say, "Take it out."
- The child removes the 10-block.
- Ask, "Why did you do that?"
- The child answers, "It is too big for the 9-box."
- Children transfer the blocks: they start on the left with the 9, then 1 + 8, 2 + 7, 3 + 6, 4 + 5 . . . 8 + 1.
- They will have one 9-block left over at the end.

The 8-Box

- Place the 8-box next to the 9-box and carry out the same experiment.

The 7-Box

- Place the 7-box next to the 8-box. The children can play familiar games but need not memorize the facts.

13. Adding 1 to a Number--The Climbing 1

Adding 1 to a Number Results in Reaching the Next Higher Number

- Scatter the blocks on the table.
- Begin a stair by placing blocks 1, 2, and 3 in the left corner of the 10-box.
- Place an extra 1-block on the first step and show that 1 and 1 equal 2.
- Call on a child to move the block up one step.
- Elicit that 2 and 1 equal 3.

* Call on another child to move the block up one step.

- The child states, "3 and 1 equal 4."
- Have the block climb the stair until it reaches the last fact, 9 and 1 is 10.

The Climbing Cat (+1, −1) (not illustrated)

- Present the stair in the 10-box.
- Place a tiny cat or an eraser on one of the steps, perhaps 4.
- Now place a screen in front of the stair.
- Say, "I moved the cat up one step. Where is it?"
- A child answers, "On the 5-block."
- Remove the screen so the children can verify the answer.

14. Adding 2 to a Number--The Climbing 2

Adding 2 to a Number Results in Reaching the Next Higher Odd or Even Number

- Have the children build a stair in the 10-box.
- Demonstrate the climbing 2 by placing the extra 2-block on the first step.
- Elicit that "1 and 2 equal 3."
- The child who makes the next move says, "2 and 2 is 4."

* The next child puts the 2-block on 3.

- The child states, "3 and 2 equal 5."
- The block climbs the stair until the last fact: 8 and 2 is 10.

Building a Stair with a Step of 2 (not illustrated)

- Scatter blocks 1 to 10 on the table.
- Ask a child to build a stair with a step of 2.
- It will be possible to build two different stairs.
- The child who starts with 2 builds 2, 4, 6, 8, 10.
- The child who starts with 1 builds 1, 3, 5, 7, 9.
- Also play the Climbing Cat game.

Checking Up: Level II

Children Talk about Their Discoveries

On Level II the children learn how to put into words the discoveries they have been making with the materials. They have learned the name for each number block by counting up to its place in the sequence or by counting its units. Now when they put two blocks together, they can tell us what they equal in their own words. They soon learn to express these facts in the language of mathematics. On Level II they should be able to do the following tasks:

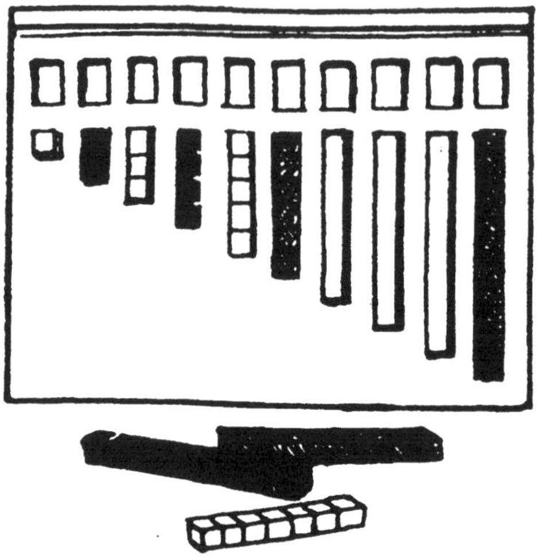

The Counting Board

- To count from one to ten
- To enumerate objects
- To give each number block its name
- To find the number block that is described by the words *one bigger than, one smaller than, comes after a certain block, comes before, comes between,* and *is equal to*

The Pattern Boards and Cubes

- To give each number pattern its name
- To understand the characteristics of odd and even numbers
- To name the number that will result when any of the following acts are demonstrated with cubes:
  - Adding 1 to a number
  - Subtracting 1 from a number
  - Adding 2 to a number
  - Subtracting 2 from a number
  - Adding or subtracting 0

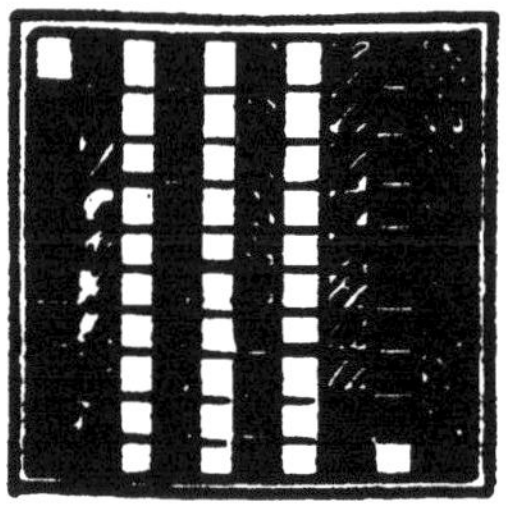

The 10-Box

- To name the block combinations that fill the 10-box
- To know the combinations without the materials
- To demonstrate the facts that result from adding 1 or adding 2 in the stair

Number Boxes 1 to 9

- To fill the number boxes 1 to 9 with blocks (children need not master all the combinations yet).

The Counting Board: Level III

Introducing Number Symbols

Goals

- To put the number symbols or numerals in place from 1 to 10 and to be able to name each one
- To know the numeral for each number block
- To know the meaning of + and = and their names, *plus* and *equal*
- To be able to read an equation in addition
- To know that the symbol 0 stands for no block at all and is named *zero*

Introduction and Naming the Numerals

Throughout the experiments on Level III the number guide should be attached to the top of the counting board. It acts as a guide when the children are learning to put each number marker in its proper place above the number block with the corresponding number of units.

As children match the numeral on the number marker with the corresponding symbol on the guide, their attention is focused on the outline, or configuration, of the numeral. To find the place of a certain numeral in the series from 1 to 10 the children need only recognize the matching figure on the number guide. A black line at the bottom of each number marker indicates the right position for that numeral. Thus, the children may turn a number such as 3 around in their hands before putting it into its place. They become well acquainted with its correct position before they learn to write it. This helps prevent reversals.

Number Symbols

A numeral, like a letter, is a written configuration of lines, a symbol: 1, 2, 3, 4, 5, 6, 7, 8, 9, 10. Each numeral stands for an amount or a number of objects. In the Structural Arithmetic approach, children work with amounts in the form of number blocks and cube patterns on Level I. Now they will learn which symbol stands for each different amount. It would be most exact to say, "Tell me the *number* of children and I will write the *numeral* on the board." However, the numerals in this approach have taken on all the qualities of the numbers. Moreover, all of us use the word *number* most of the time; for instance, we say *telephone number, route number,* or *social security number,* so Structural Arithmetic follows suit and says *number marker* and *number guide.* The experiments are devised in ways that make this clear: each configuration is a symbol that stands for a number of cubes, number of children, or pennies, or a block of a certain size.

Self-Teaching Aspects of the Materials

One of the most important principles built into these materials is their arrangement, which keeps many different relationships in view and in the consciousness of the children at the same time. Attaching the number guide and using the number markers converts the counting board into a self-teaching device. The children can discover the name of each numeral and which numeral is meant by a certain name. Some other relationships are as follows: for example, a child picks up the numeral 4, which for a moment might be all alone and meaningless. Then she or he spots its place between 3 and 5 and the ordinal concept is brought to mind as the child realizes 4 is now in fourth place. At the same time, he or she cannot miss the cardinal concept supplied by the 4-block. The figure 4 takes on the meaning of the number 4, an amount much smaller than 10, about half way in size, but belonging among the little ones like 1, 2, 3, and 5.

Testing and Remedial Work

The counting board, with its removable symbols, can be used for testing in remedial work. If the numbers are presented one at a time, the teacher can see whether the children know the relative position of each, that is, do they know that 9 comes just before 10, or must they count up from 1? The Snake Game, in which a symbol is selected, read, and the corresponding block removed from the board for the building of a snake of blocks, can be played by two competing teams. The game is fun and gives the children doing remedial work a stronger foundation in the meaning behind each numeral.

To these numerals the teacher can add zero. The children are told that this means "no block." The 0-marker is then turned face down and placed with the other number markers. It is very dramatic, indeed, when a child draws 0. If it is too crushing, once a child has understood the meaning, she or he can be given another, more rewarding, turn. The difference between getting nothing and the chance to have a number block seems very great to children.

When the children have learned the symbol that stands for each number block, they are ready to learn to record the number combinations they have been discovering in the experiments with the materials.

The children learn that the plus sign stands for an operation or an action performed with the materials. In the first games, we say that the plus sign means that the two blocks hidden beneath a box are connected.

The equal sign is presented after children know the plus sign well. It stands for "equal" or "is just the same size as." When two blocks are put together they equal another block. The two lines that make up the equal sign are "just the same size as" each other. This explanation helps children remember what the sign means.

They are now ready to use both signs in an equation.

1. Introducing Number Symbols 1 to 10

- Display the counting board filled with blocks.
- Attach the number guide at the top of the board.
- Place the number markers in an empty 10-box.
- * Point to the figure 1 on the number guide.
- Say, "This is one."
- Ask, "Who can find number 1? Put it in its place."
- Pass the tray of number markers to a child.
- The child should recognize the matching symbol on the number guide and put the marker in its place.
- Since the children know the 1-block is called "one," they understand that the figure 1 has the same name.
- Continue until all the numerals have been named and each is in its correct place in the sequence 1 to 10.

Finding a Named Number--A Calling Back Game (not illustrated)

- Display the counting board and blocks with the number guide in place.
- Place the number markers face up in a tray.
- Say, "Calling back 3!" A child puts 3 in place.
- Continue until all the number markers are back in place.

2. Letting the Numbers Call Back the Blocks

- Display the counting board with the number guide in place.
- Scatter blocks 1 to 10 on the table.
- Put the number markers face up in a tray.
- * Say, "Pick a number; where does it go?"
- A child selects 3 and puts it in its place.
- Ask, "What is it?" The child answers, "3."
- Say, "Put back the block it tells you to."
- Continue until all the number markers and their blocks are in place.

The Silent Teacher Game (not illustrated)

- Display the counting board with the number guide in place.
- Pass out blocks to the children.
- Keep the number markers for yourself.
- Explain that in the Silent Teacher Game, no one can say a word, not even the teacher. You will put a number in place, and without a word, the child with that block puts it back.
- Continue in silence until all the blocks are back in place.

3. Matching Blocks with Numbers--Introducing Zero

The Snake Game

- Explain the Snake Game as follows: The group will be divided into two teams. The team members take turns picking a number that tells them which block to get for their team's snake.
- Display the counting board with blocks and number guide in place.
- Put the number markers 1 to 10 face down on a tray.
- Hold up the marker with 0 (zero).
- Explain that in this game picking zero means you get "no block at all."
- Place 0 with the other numbers on the tray.
- Pass the tray of number markers to a child.

* Say, "Choose a number; put it back" (in the board). The child chooses 9 and puts it in the counting board.

- Say, "You can take that block for your snake."
- The child adds the 9-block to his or her team's snake.
- Assist the children in forming their snakes of blocks.

4. Checking Up: Matching Numbers to Blocks without the Guide

- Display the empty counting board with the number guide removed.
- Place the blocks on a table across the room.
- Place the number markers face down in a tray.
- Pass the tray to a child.

* Say, "Pick a number; go to the table and get that block."

- The child puts the block and number marker into the counting board.
- Continue until all are in place.

Note: If a child does not recognize a number, have him or her count up the spaces in the counting board to where he or she thinks it "lives."

The Silent Teacher Game

- Display the counting board without the guide. Play the Silent Teacher Game on page 44.

5. Understanding That One Number Can Be Made of Two Smaller Numbers

Note: Children must first understand the relationship between two numbers and their sum. What do we mean when we say two numbers "make" another number, or "equal" a number?

- Display the counting board filled with blocks.
- Scatter another set of blocks 1 to 10 on the table.
- * Remove the 5-block from the board.
- Put the number marker 5 in its place.
- Ask, "Can you find two blocks that together make 5?" (point to the empty 5-groove).
- Call on a child, who puts in, perhaps, 2 and 3.
- Ask, "What are they?"
- The child answers, "2 and 3."
- Ask, "Can you tell that story?"
- The child answers, "2 and 3 make 5."
- Continue making other combinations.

Note: This is another way to show the combinations that make 10. Remove the 10-block and put in the 10-marker. Have children fill the groove with combinations such as 9 and 1. Use an extra 5 block to make 5 and 5.

6. The First Number in the Equation

- Display the empty counting board.
- Scatter blocks 1 to 10 on the table.
- Keep the number markers for yourself.
- Put another set of blocks 1 to 10 on a nearby table.
- Show the children your hiding box.
- Explain that you will hide a block under the box and give one clue for them to figure it out.
- Say, "Close your eyes" and hide the 4-block.
- Say, "Open your eyes."
- * Say, "I have hidden a block under this box. The number gives you the clue" (place number marker 4 on top of the box).
- Ask, "Can you get the twin of my block and put it here?" Point to the child's side of the box.
- A child says "4" and gets the 4-block.
- Encourage children to look under the box for the twin 4-block.
- The child puts the symbol and 4-block into the board.
- Let the child be the next teacher.

7. Introducing the Plus Sign

- Display the counting board with blocks and the number guide in place.
- Keep the number markers for yourself.
- Remove blocks 1 to 6 and scatter them on the table.
- Explain that you will hide two blocks under the box.
- Say, "Close your eyes" and hide the 3- and 2-blocks.
- Say, "Open your eyes."
* Say, "I've hidden two blocks under this box. These numbers tell you what they are."
- Place markers to show 3 + 2 on top of the box.
- Ask, "Who can get the blocks and put them here?" (point).
- Call on a child, who reads, "3 and 2," and places the blocks where the teacher points.
- Ask, "Is there another way to say this?"
- Some children may know "3 plus 2."
- If the word *plus* is new, teach it to them.
- Continue with other blocks and numbers.

Note: Work with the plus sign until the children feel secure with it. Only later, introduce the equal sign.

8. Introducing the Equal Sign

Reading an Equation

- Display the counting board with blocks, number markers, and number guide in place.
- Set up the number markers 3 + 2.
- Have a child read the number markers and place the blocks in the same order (3 + 2).
* Ask, "Can you find a block just as big as 3 and 2?"
- The child gets the 5-block and says, "5."
- Display the equal sign. Say, "This sign says 'equals.'"
- Ask, "Who can read this example?"
- A child reads, "3 plus 2 equals."
- Say, "Excellent! Get the answer and put it here" (point to the end of the example 3 + 2 =).
- The child gets the symbol 5 from the board and puts it at the end of the equation.
- Another child reads, "3 plus 2 equals 5."
- Have other children read this equation.
- Continue with other blocks and numbers.

Note: Prevent the children from mixing up the plus sign and equal sign. Wait until they are secure with the plus sign before you introduce the equal sign. Explain that the two lines in the equal sign are just as big as each other, which is what the sign means.

Pattern Boards: Level III

Introduction

Goals

- To know the number that stands for each pattern
- To know the meaning of +, −, and =
- To be able to act out a subtraction word problem
- To be able to read and understand an equation
- To be able to use the plus or minus sign in an equation

On Level II the children learned the number name for each pattern and the name of each digit in the counting board. It should be easy for them to connect the correct numeral or number symbol to each pattern. They can use counting for verification.

Using the Plus and Minus Signs

The plus sign was introduced as meaning "and" when two blocks were added together, for example, three and two (3 + 2). The meaning of this sign is now extended. The children will realize it also stands for another action. Reading the expression 6 + 1 will mean not only the addition of 1 cube to the 6-pattern but also "going up one" in the number series from the 6-board to the 7-board.

Introducing the Minus Sign As the Opposite of the Plus Sign. On Level II the children became familiar with adding 1 cube to a pattern and then with the opposite action of taking 1 cube away. They added 2 cubes and then subtracted 2 cubes, moving in the reverse direction. On Level III children will locate a prize (a sticker) after reading the expression 6 − 1. They will figure out that this can also mean "go down" one board to the 5-board. They will then record the rest of the equation with the proper symbols: 6 − 1 = 5.

Combinations: The Doubles and the Neighbors

These are among the most difficult combinations for children to remember by rote. When they can visualize the patterns, they will find them easier to figure out. There are two sets of combinations: the even pattern boards show the doubles; the odd pattern boards show the sums of consecutive numbers (neighbors).

1 +0	2 +1	3 +2	4 +3	5 +4
1 +1	2 +2	3 +3	4 +4	5 +5

Making Up Word Problems and Acting Them Out

Having children compose their own addition and subtraction word problems is by far the best way to get them to understand the word problems they will later listen to or read. It is easiest for them when you assign a cube pattern; they then remove, or subtract, some cubes while telling the story. This experiment makes clear that they must begin with the larger amount and end up with a smaller amount. They follow this experiment by demonstrating a story the teacher is telling.

Recording an Operation: Understanding the Opposite Meanings of the Plus and Minus Signs

In this final experiment, the children are shown both signs, plus and minus, along with the numbers. The teacher then takes a number pattern, a child records it with a number marker, and then the teacher "does something" to it behind a screen. When the pattern board reappears, the children see that it has either been diminished by a certain amount or increased. By looking at the changed pattern, they can infer the action that the teacher carried out behind the screen. If the number is less, they record the action with a minus sign; if it is more, with a plus sign. Thus, the opposite meaning of each sign is established in their minds. Setting up equations in the Screen Game prepares them for visualizing the action that takes place in a word problem. This will enable them to extract the basic addition or subtraction equation from the text, a most difficult task, especially in algebra. Thus, the foundation for solving problems in algebra can be prepared at this level.

1. The Numbers Call Back the Pattern Boards

- Pass out the pattern boards to the children.
- Keep the number markers for yourself.
- Introduce the number symbols in sequence from 1 to 10.
- Hold up number marker 1.
- Say, "Calling back the 1-board; where does it live?"
- The child with the 1-board puts the board in first place with number 1 above it.

* Hold up number marker 2. Say, "Calling back this pattern board."

- Ask, "What is it? Where does it live?"
- The child with the 2-board announces that the name is "2" and puts it in place.
- Continue until each pattern board is in place with its number above it.

Note: The children have already learned the names of the number symbols (*two* is the name for figure *2*). Since they know the name of each pattern, it is easy for them to connect each number to its pattern. Counting should not be necessary, but it may be used for verification.

Variation: Try putting boards on a table across the room. Let the children draw a number marker and get the board.

2. Reading the Plus Sign

- Buy small stickers for the children to win.
- Place the empty pattern boards in sequence.
- Have available the number markers 1 to 10 and the plus sign.
- When the children's eyes are closed, hide a sticker beneath the 7-board.
- Show 6 + 1 with the number markers.
- Have the children open their eyes and look.

* Say, "Watch! This tells you where the sticker is hidden" (point to the markers).

- Call on a child. Say, "Please read it."
- The child answers, "6 and 1."
- Ask, "Which pattern is that?"
- The child answers, "7" and looks under the board to find the hidden sticker.
- Continue the game using other patterns.

Note: You can also explain that +1 means "go up one" board. The child puts a finger on the 6-board and moves up one to the 7-board.

3. Introducing the Minus Sign

- Have the children fill the pattern boards with cubes and place them in sequence.
- Tell the children to close their eyes.
- Hide a sticker under the 5-board.
- Arrange the number markers to read 6 − 1.
- Have the children open their eyes.

* Say, "This message tells you where to find the sticker."

- Explain, "This sign says, 'Take away.'" (Point to the minus sign. You may prefer to say "minus.")
- Ask, "Who can read this?"
- A child answers, "6 take away 1 . . . 5." (The child may want to take 1 cube from the 6-pattern to see the 5-pattern emerge.)
- Another child looks under the 5-board and finds the sticker beneath it.
- Explain that −1 can also mean "go down one" board. You touch the 6-board and move down to the 5-board.
- Continue giving other number messages.

Note: Contrast −1 with +1 and −2 with +2. Don't forget to use −0 and +0. You may say either "minus" or "take away."

4. Reading and Understanding a Subtraction Equation

- Put the filled pattern boards in sequence on the strip.
- Arrange the number markers to read 8 − 1 = __.

* Ask, "Who can read this example?"

- A child answers, "8 take away 1 equals."
- Ask, "Which board is that?"
- Another child answers, "7."
- Say, "Take the number marker and put it in the equation."
- A child does so and reads, "8 take away 1 equals 7."

Play the Dumping Out Game as described on page 12.

- A child dumps out cubes from the 7-board.
- Continue with easy subtraction examples. Use +0, −0; +1, −1; and +2, −2.

. Combinations: The Doubles

- Arrange the even pattern boards on the cardboard strip.
- Place a pencil lengthwise down the center of each board in turn, showing the doubles: 1 + 1, 2 + 2, 3 + 3, 4 + 4, and 5 + 5.
- Recite each combination as you point to that board.
- Have available two paper cups and cubes.
- To demonstrate the 6-pattern, fill each cup, one with 3 light-colored cubes and the other with 3 dark-colored cubes.
- Ask, "What's in this cup?"
- Dump out the cup and scatter the cubes.
- A child answers, "3."
- Ask, "What's in this cup?"
- Dump out the other cup.
- A child answers, "3."
- Ask, "How many altogether?"
- A child answers, "6," puts the cubes into the 6-board, and places the number marker 6 above it.
- Show the other doubles: 1 + 1, 2 + 2, 4 + 4, and 5 + 5.

Note: The pattern board brings about a perceptual unification of the two parts.

6. Combinations: The Neighbors (Sums of Consecutive Numbers)

- Arrange the odd pattern boards on the cardboard strip.
- Place a pencil lengthwise down the center of each board in turn to show the combinations 1 + 0, 2 + 1, 3 + 2, 4 + 3, and 5 + 4.
- Have available two paper cups and cubes.
- To demonstrate the 7-pattern put 4 dark-colored cubes in one cup and 3 light-colored cubes in the other cup.
- Ask, "What's in this cup?"
- Dump out the cubes.
- A child answers, "4."
- Ask, "What's in this cup?"
- Dump out the other cup.
- A child answers, "3."
- Ask, "How many altogether?"
- A child answers, "7," and puts the cubes into the 7-board, arranging them by color to show rows of 4 and 3 cubes.
- The child places number marker 7 above the board.
- Show the other combinations: 1 + 0, 2 + 1, 3 + 2, and 5 + 4.

Note: Each part differs from the whole pattern. When they are put together, they become parts of an organized whole.

7. Doubles and Neighbors: The Missing Addend

- Keep pattern boards 1 to 10 on the cardboard strip.
- Keep a supply of cubes for yourself.
- Place each number marker above its pattern board.
- Tell the children to close their eyes.
- Think of either a doubles combination or the sum of consecutive numbers, for example, the 7-pattern.
- Put 4 cubes in one hand and 3 cubes in the other.
- * When the children have opened their eyes, say, "I have 7 cubes in all. Point to the 7-board" (a child points).
- Say, "In one hand I have 4 cubes" (drop them).
- Ask, "How many do I have in the other hand?"
- The child looks at the 7-board and answers, "3."
- Open your hand to show the missing addend.
- The child puts the cubes into the 7-board, filling the rows of 4 and 3 cubes.
- This is excellent practice for these hard combinations.

8. Checking Up: Doubles and Neighbors

- Put the filled pattern boards in sequence. Use rows of dark- and light-colored cubes as illustrated.
- Print the following combinations on ten cards: 1 + 0, 1 + 1, 2 + 1, 2 + 2, 3 + 2, 3 + 3, 4 + 3, 4 + 4, 5 + 4, and 5 + 5.
- Give each child an opportunity to place each card below the appropriate board.
- Collect the cards; leave the boards filled with cubes for the next game.

The Sticker Game

- Explain that while the children's eyes are closed, you will hide a sticker underneath one of the pattern boards.
- Hide a sticker beneath the 6-board.
- * Say, "Open. I hid a sticker under a board. This card tells you which board."
- Hand the child the card (3 + 3).
- Say, "Please read this card."
- A child reads, "3 and 3. It's 6."
- The child looks beneath the 6-board and finds the hidden sticker.
- Hide stickers under other boards.

Word Problems: Telling Subtraction Stories

Explain to the children that they are going to tell subtraction stories and act them out with their cube patterns.
Give each child a board filled with cubes.
Show a demonstration of your own.
Say, "Once I had 8 kittens." (Use the 8-board filled with cubes.)
"I gave 4 kittens away" (remove 4 cubes).
Conclude, "I had 4 kittens left."
Call on children to make up their own stories.
The child with the 6-board says, "My dog had 6 puppies. I gave 3 away" (removes 3 cubes); "I have 3 left for myself."
Call on each child to act out a story with his or her number pattern.

Note: In acting out stories the children discover the different roles that the numbers play. The storyteller begins with a total (pattern board) and must mentally take it apart into two parts. One part is subtracted, while the other remains. The board is made so that the original amount is visible after the subtraction has taken place. Thus, the relationships between parts and the whole can still be seen.

Comprehending Oral Subtraction Stories

Note: This game prepares children for understanding word problems.

Explain that you will tell a story and then ask who has the pattern board that can be used to show it.
Give the children pattern boards filled with cubes.
Tell a subtraction story such as "I had 10 cookies. I ate 5. How many were left?"
Ask, "Who can show that story?"
The child who has the 10-board raises a hand and says, "I can."
The child says, "I had 10 cookies. I ate 5 cookies" (removes 5 cubes).
The child concludes, "I have 5 cookies left."

Note: When the children listen to an oral subtraction problem, they hear two numbers mentioned (in this case, 10 and 5). They often do not know which number "comes first" when they write the example. Encourage the children who need it to demonstrate this problem. If they start with the 5-board, they will discover they cannot take 10 away from 5. The words will suddenly take on meaning for them. Later the children record these stories with equations.

11. Stating a Subtraction Problem in Words

- Set up a cardboard screen.
- Keep the cubes and pattern boards for yourself.
- Select an odd pattern board such as 7; show 4 + 3 by using rows of light- and dark-colored cubes.
- * Ask, "How much is this?" (point to board).
- A child answers, "7."
- Now, behind the screen, remove the row of 3 cubes.
- Reveal the 7-board with the row of 3 cubes missing.
- Say, "What did I do? Tell me! "
- A child answers, "You took 3 cubes away."
- Ask, "What is left?"
- A child answers, "4 cubes."
- Work with other odd patterns to practice sums of consecutive numbers.

Note: In their minds, the children visualize an action, that is, what the teacher is doing. They learn to think in terms of materials and actions. Putting actions into words lays the foundation for the comprehension of word problems.

12. Recording the Operation--Did I Add or Did I Take Away?

- Set up the vertical screen.
- Divide a blank 3" x 5" card into three sections to hold three number markers.
- Keep the pattern boards and cubes for yourself.
- Place the numbers face up on a tray.
- Display the markers with + and − . Review the meanings of the signs.
- * Build the 7-pattern (showing the combination 4 + 3).
- Ask, "How much is this?" (point to the 7-pattern).
- Say, "Get that number, please!"
- A child places the 7-marker in first place on the card.
- Move the 7-board behind the screen and remove 3 cubes.
- Reveal the 7-pattern with the row of 3 cubes empty.
- Ask, "What did I do? Show me by getting the right sign. Did I add or did I take away?"
- A child picks the minus sign and places it next to the 7.
- Now ask, "How many cubes did I take away?"
- The child records the amount with the 3-marker.
- Ask, "Who can read this?"
- A child answers, "7 take away 3" or "7 minus 3."
- Continue with other patterns, adding as well as subtracting cubes.

Note: It is important that children master the opposite meanings of the plus and minus signs before they complete the equation with the equal sign and the answer.

The 10-Box and Number Boxes 1 to 9: Level III

Introduction

Goals

- To record from memory addition facts with the sum of 10
- To record the related subtraction facts
- To demonstrate and record the equation from hearing an addition story
- To demonstrate and record the equation from hearing a subtraction story

Recording the Story of 10

On Level II, the children understood and learned to state the addition facts that they found with the Structural Arithmetic materials, saying, "6 and 4 are as big as 10." In the counting board experiments on Level III they learned to use the plus sign and the equal sign to set up an equation. Now they will learn to record such a fact using the proper signs and symbols: 6 + 4 = 10.

The teacher now places the 10-box and the number markers before the children and inserts the first two addends. The children name these two addends, for example, six and four, and record them with the plus sign: 6 + 4. When the children say aloud the next step in the equation, they are dictating to themselves the symbols that come next. "Equal ten" is recorded by putting those symbols last: 6 + 4 = 10.

The teacher should make certain that the children understand each step.

Reading Equations

Children make up equations with the comprehension that comes from recording the combinations of blocks as they place them in the box. This enables them to understand an equation when they read it.

The Missing Addend Form. Many children find this form (6 + __ = 10) difficult to comprehend when they first meet it in print because they have no understanding of what it means. When children put the 10-block in the box and place the 6-block next to it, they find it natural to say "6 needs what to make 10?" They can record what they have said with the number markers: 6 + __ = 10. The missing number symbol is the number of the missing block. This helps children comprehend an equation in which one addend is missing.

Recording a Subtraction Equation. Subtracting one of two addends from the 10-box leaves the other. Children learn to put this into words in preparation for recording subtraction facts. They are then able to dictate to themselves the symbols and signs as they record a fact such as "10 take away 6 leaves 4."

The test of their understanding comes in their ability to record the equation that is derived from an oral story that the teacher demonstrates behind a screen. They listen intently and place the number markers and symbols in the form of an equation. They check their symbols later against the blocks the teacher reveals.

Recording an Addition Equation. The children have learned to put together an addition equation from working with blocks or number patterns. Can they record the equation when they hear a word problem in addition? The teacher can help them visualize the action by telling the story as she hides the corresponding blocks beneath a box or scarf. Later, the children check the equation by looking at the hidden blocks.

Recording the Number Stories: Boxes 1 to 10. The stories recorded from the combinations of blocks in the number boxes are another way to look at the fifty-five number combinations in addition that all children need to know.

10	9	8	7	6	5	4	3	2	1	
+0	+1	+2	+3	+4	+5	+6	+7	+8	+9	10-box
9	8	7	6	5	4	3	2	1		
+0	+1	+2	+3	+4	+5	+6	+7	+8		9-box
8	7	6	5	4	3	2	1			
+0	+1	+2	+3	+4	+5	+6	+7			8-box
7	6	5	4	3	2	1				
+0	+1	+2	+3	+4	+5	+6				7-box
6	5	4	3	2	1					
+0	+1	+2	+3	+4	+5					6-box
5	4	3	2	1						
+0	+1	+2	+3	+4						5-box
4	3	2	1							
+0	+1	+2	+3							4-box
3	2	1								
+0	+1	+2								3-box
2	1									
+0	+1									2-box
1										
+0										1-box

1. Recording the Story of 10

- Empty the 10-box, scattering the blocks on the table.
- Place number markers 1 to 10, + −, =, and 0 face up on a tray.
- Place a combination of blocks such as 7 and 3 in the 10-box.

* Ask, "Who can start to record this equation with numbers?"

- A child places the markers for 7 + 3 on the table and says, "7 plus 3."
- Ask, "What sign comes next?"
- A child answers, "Equals," and adds the sign to the equation.
- Ask, "Now, what comes last?"
- A child answers, "10!" and adds the number marker to the equation.
- A child reads the equation and the other children check it.

2. Finding the Missing Addend

- On ten cards, write equations with the addends missing. The space for the missing addend should be large enough for a number marker. 10 + _ = 10, 9 + _ = 10, 8 + _ = 10, . . . 1 + _ = 10.
- Turn the pile of cards face down.
- Scatter the blocks from the 10-box on the table.
- Put the number markers 1 to 10 and 0 face up on a tray.

* Say, "Pick a card; read it. Get your blocks. Find the missing number."

- A child reads, "7 plus what equals 10?"
- The child gets number marker 3, puts it on the card, and keeps both blocks and card.
- Give turns to other children.

3. Recording the Climbing 1

- Scatter the blocks from the 10-box on the table.
- Build the first steps of the stair: 1, 2, 3, 4, 5.
- Have the children recite with you as the 1-block climbs the stair, "1 + 1 = 2, 2 + 1 = 3, 3 + 1 = 4, 4 + 1 = 5."
- Explain that they will record the rest of the climbing 1.
- Call on a child to record the next step with the number markers.

* Move the 1-cube on top of 5.

- Ask, "5 plus 1 equals?"
- The child records, 5 + 1 = .
- Another child answers, "6," and places the 6-block next to blocks 5 and 1.
- Respond, "Right! Now, put 6 in the equation."
- Continue to the last equation, 9 + 1 = 10.

4. Adding to 1

- Scatter the blocks from the 10-box on the table.
- Select a block such as 6 and stand it up.

* Say, "I'm going to give you the answer first."

- Place the 1-cube next to the 6-block.
- Say, "Now, what do you have to add to 1 to get as high as 6?"
- A child says, "5" and puts a 5-block on top of the 1-cube.
- Ask, "Can you record this?"
- The child records with number markers: 1 + 5 = 6.
- Ask, "Who wants to proofread this equation?" (Children like this!)

5. Subtracting from 10

The Scarf Game

- Put the blocks from the 10-box on a side table near your chair.
- Cover the empty 10-box with a scarf.
- Explain that you are going to hide in the 10-box two blocks that make 10.
- Say, "Close your eyes."
- Place 4 and 6 in the 10-box under the scarf.
- * Say, "Open your eyes! I hid two blocks under here that together make 10."
- Announce, "If Ann takes 6 away, what will be left?"
- Call on a child to answer "4."
- Tell Ann, "OK, Ann. Take 6 away."
- Another child removes the scarf and reveals the remainder, 4.
- Have a pupil state, "10 minus 6 leaves 4" or "10 take away 6 equals 4."
- Now, show the related addition fact by having the 6-block rejoin the 4 to make 10 again.

Subtracting from Boxes 1 to 9 (not illustrated)

- Review the combinations in each box before starting subtraction.
- Hide the box beneath the scarf and play the Scarf Game described above.

6. Word Problems in Subtraction--Recording the Equations

- Place the number blocks on the table.
- Put number markers 1 to 10, 0, +, −, and = face up in a tray.
- Explain that you will hide two blocks under a box and then tell a story about them. The children will record the story with the number markers.
- When the children's eyes are closed, hide blocks 9 and 1 under the box.
- Say, "Open your eyes."
- * Say, "I had 10 bananas. I ate 1 of them" (withdraw the 1-cube from the box).
- Ask, "How many are left?"
- Ask a child, "Would you record that story with numbers?"
- The child selects number markers and dictates to herself as she records, 10 − 1 = 9.
- Have another child read the equation and lift up the box to show that the 9-block is left.
- Put the 1-block back with the 9-block to show the corresponding addition fact.
- Do other problems.

7. Word Problems in Addition--Recording the Equations

- Place the number blocks on the table.
- Put number markers 1 to 10, 0, +, −, and = face up in a tray.
- Explain that you will hide two blocks under a box and then tell a story about them. The children will record the story with the number markers.
- When the children's eyes are closed, hide blocks 8 and 2 under the box.
- Say, "Open your eyes."
- Say, "I had 8 bananas. I went into a store and got 2 more. How many did I have then?"
- Ask a child, "Would you record that story with numbers?"
- The child selects number markers and dictates as he records, "8 + 2 = 10."
- Have another child read the equation and lift up the box to reveal the blocks.
- Continue with other stories.

8. Studying Combinations That Make 5 or 6

- Select two number boxes, such as the 5- and 6-boxes.
- Place them side by side.
- Scatter blocks on the table from the counting board and the 10-box.
- Write the following examples on cards large enough to accommodate the number markers: 5 + 0 = , 4 + 1 = , 3 + 2 = , 2 + 3 = , 1 + 4 = ; and 6 + 0 = , 5 + 1 = , 4 + 2 = , 3 + 3 = , 2 + 4 = , 1 + 5 = . Put the cards in a pile.
- Say, "Pick a card. Get the blocks it tells you to."
- Ask, "Which box will they fit in--the 6-box or the 5-box?"
- The child reads, "3 plus 2 equals 5" and gets these blocks to put into the 5-box.
- Do the same with other pairs of number boxes.

Note: Since there are not enough blocks to fill both boxes, one of the boxes will be the "winner."

Checking Up: Level III

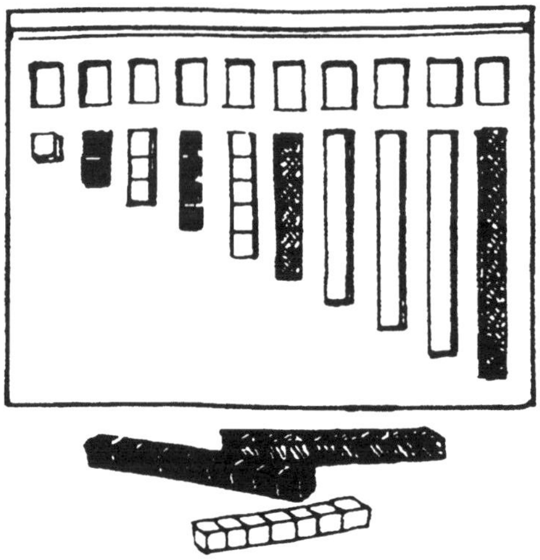

On Level III, the number symbols 1 to 10 and 0 as well as the signs +, −, and = have been introduced. The children should be able to do the following tasks:

The Counting Board
- To name each number symbol or numeral from 1 to 10
- To put the numbers in sequence from 1 to 10
- To know the number for each block
- To know the meaning of the plus sign
- To know that 0 stands for no block

The Pattern Boards and Cubes
- To know the number for each pattern
- To know the meaning of the signs +, −, and =
- To be able to act out a subtraction word problem with cubes
- To read and understand an equation
- To use + or − and = to form an equation

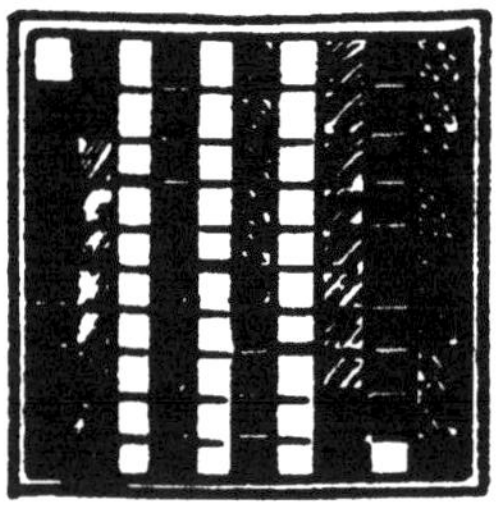

The 10-Box
- To record addition facts with a sum of 10 from memory
- To record the related subtraction facts
- To demonstrate and record an equation from hearing an addition story
- To demonstrate and record an equation from hearing a subtraction story

Number Boxes 1 to 9
- The same goals as for the 10-box but without complete mastery of the facts

SCOPE AND SEQUENCE

LEVEL I
Actions and the Formation of Concepts

	COUNTING BOARD				PATTERN BOARDS					10-BOX			
Page:	6	7	8	9	11	12	13	14	15	17	18	19	20
Cardinal numbers	●	●	●	●	●	●	●	●	●	●	●	●	●
Sequence 1 to 10		●				●			●			●	●
Equality			●		●	●	●	●	●	●	●		
Number names													
10-box facts										●	●		
Concept of 0										●			
Adding 1													
Subtracting 1													
Odd and even nos.													
Adding 2													
Subtracting 2													
9-box facts													
8-box facts													
7-box facts													
6-box facts													
5-box facts													
4-box facts													
3-box facts													
2-box facts													
1-box facts													
Numerals													
+ sign													
= sign													
− sign													
Word problems													

LEVEL II
Number Names and Language

	COUNTING BOARD					PATTERN BOARDS					10-BOX						
Page:	23	24	25	26	27	29	30	31	32	33	35	36	37	38	39	40	41
Cardinal numbers	●	●	●	●	●	●	●	●	●	●	●	●	●	●	●	●	●
Sequence 1 to 10	●	●	●	●	●	●	●	●	●	●		●	●			●	●
Equality						●					●	●	●	●	●	●	●
Number names	●	●	●	●	●	●	●	●	●	●	●	●	●	●	●	●	●
10-box facts											●	●	●	●	●	●	
Concept of 0							●		●	●	●	●	●	●	●	●	
Adding 1			●	●	●		●			●	●	●	●	●	●	●	●
Subtracting 1			●	●	●		●			●							●
Odd and even nos.								●	●	●						●	
Adding 2									●	●	●	●	●	●	●	●	●
Subtracting 2									●	●							●
9-box facts																●	
8-box facts																●	
7-box facts																●	
6-box facts																●	
5-box facts																●	
4-box facts																●	
3-box facts																●	
2-box facts																●	
1-box facts																●	
Numerals																	
+ sign																	
= sign																	
− sign																	
Word problems																	

LEVEL III
Numerals; Understanding and Setting Up Equations

	COUNTING BOARD				PATTERN BOARDS						NUMBER BOXES 1-10			
Page:	44	45	46	47	49	50	51	52	53	54	56	57	58	59
Cardinal numbers	●	●	●	●	●	●	●	●	●	●	●	●	●	●
Sequence 1 to 10	●	●			●	●	●	●				●		
Equality			●	●	●	●	●	●	●	●	●	●	●	●
Number names	●	●	●	●	●	●	●	●	●	●	●	●	●	●
10-box facts			●				●	●	●	●	●	●	●	●
Concept of 0		●	●			●	●	●			●		●	●
Adding 1			●		●	●	●	●		●	●	●	●	●
Subtracting 1						●				●			●	
Odd and even nos.					●	●	●	●	●	●				
Adding 2			●	●		●	●	●	●	●	●			●
Subtracting 2						●			●	●			●	
9-box facts							●	●	●	●			●	●
8-box facts							●	●	●	●			●	●
7-box facts							●	●	●	●			●	●
6-box facts							●	●	●	●			●	●
5-box facts							●	●	●	●			●	●
4-box facts							●	●	●	●			●	●
3-box facts							●	●	●	●			●	●
2-box facts							●	●	●	●			●	●
1-box facts							●	●	●	●			●	●
Numerals	●	●	●	●	●	●		●		●	●	●	●	●
+ sign				●	●	●		●		●	●	●		●
= sign				●		●		●		●	●	●	●	●
− sign						●				●			●	
Word problems									●	●			●	●

Made in United States
Orlando, FL
04 November 2026

90220726R00039